MW01064280

Calculator Enhancement

for Introductory Statistics

A Manual of Applications using the
Sharp EL-5200, HP-28S and HP-48S Graphing Calculators

Iris Brann Fetta, Clemson University
D.R. LaTorre, Clemson University, Consulting Editor

Saunders College Publishing
A Harcourt Brace Jovanovich College Publisher
Fort Worth Philadelphia San Diego New York Orlando Austin San Antonio
Toronto Montreal London Sydney Tokyo

Printed in the United States of America.

Fetta: CALCULATOR ENHANCEMENT FOR INTRODUCTORY STATISTICS: A MANUAL OF APPLICATIONS USING THE SHARP EL-5200, HP-28S and HP-48 GRAPHING CALCULATORS, 1/E

ISBN 0-03-092726-9

234 066 98765432

PREFACE

Careers which require some degree of mathematical ability and technical competence, including those which require a basic understanding of fundamental statistical processes, are ever-increasing. The basic concepts of probability and statistics are now presented to a mass of students from a variety of different disciplines.

Many of these students have more emotional than intellectual problems with mathematics. Often this emotional reaction by the student results in avoiding the problem by not attending class, not doing the assigned homework, and consequently, not learning the material. The resulting math anxiety is quite often a question of attitude rather than aptitude. The focus on detail and accuracy can be very unpleasant, frustrating and annoying to the non-technical student, thereby increasing math anxiety.

One of the views on how students learn mathematics is constructively. In discussing this process of making mental constructions to handle mathematical concepts, Annie and John Selden state that "The implications for mathematics teaching is that the emphasis should shift from ensuring that a student can correctly replicate what he or she has been shown to concentrating on helping her or him organize and modify her or his 'mental schemas'." (*UME Trends*, March, 1990) Frank Demana, Tom Dick, John Harvey, John Kenelly, Gary Musser, and Bert Waits report that "Mathematics teachers have long known that a good picture can be worth much more than a thousand words. A graph not only effectively captures a wealth of information about a function or relation, it also provides a visual heuristic for interpreting that information. Calculators with interactive capabilities to display graphic images are exciting new tools for teaching and learning mathematics." (*The Computing Teacher*, April, 1990)

The *Curriculum and Evaluation Standards for School Mathematics* assumes that for grades 9 through 12, "Scientific calculators with graphing capabilities will be available to all students at all times." (National Council of Teachers of Mathematics, 1989)

Statements such as these reflect the changing attitude of the mathematics community toward this new technology. Graphing calculators add a significant dimension to mathematics instruction and effectively used, should help alleviate math anxiety and increase the level of student interest in the subject matter. Demana and Waits relate that "Technology can help students think more deeply about mathematics, facilitate generalization, empower students to solve difficult problems, and furnish concrete links between geometry and algebra, algebra and statistics, and real problem situations and associated mathematical models." (*Mathematics Teacher,* January, 1990)

In an effort to incorporate this new technology, Clemson University integrated the use of graphic calculators into the undergraduate curriculum in the fall of 1989. Substantial funding for this project was provided by the Fund for Improvement of Postsecondary Education (FIPSE). We sincerely thank FIPSE for their attention to innovation and a desire to improve postsecondary education in America.

Calculator enhanced probability and statistics places more emphasis on graphing, less emphasis on actual computation, and offers hands-on experience with simulation. The student is actively involved while in the classroom. The lecture format of a class can become a more informal exploration of concepts by both instructor and students. Nothing will be gained in the total learning process by teaching students to press keys and obtain final answers. While this new technology frees the student in an introductory probability and statistics course from tedious and time-consuming calculations, major advantages are that more detailed explanation of concepts can be offered due to the relative ease of simulation and that drill and practice exercises can effectively be supplemented with realistic problems which encourage exploration and experimentation. Students in calculator enhanced probability and statistics generally comment that the programmable graphing calculator makes the course much less time consuming, and therefore, it seems easier and more interesting. Many students agree that their use of the calculator reduces anxiety over testing because careless errors are not readily made. They have found applications for the calculator to other courses, research, and off-campus jobs.

Even though the subject matter and concepts discussed in this manual could be taught using any calculator with similar performance capabilities, the specific adaptations in this work are for the Sharp EL-5200 graphic scientific calculator, the HP-28S Advanced Scientific Calculator and the HP48SX Scientific Expandable Calculator. This does not constitute an endorsement by the author, however. Other calculators such as Casio's fx-7000G, fx-7500G, fx-7700G, fx-8000G or fx-8500G or Texas Instruments' TI-81 could readily be used by students familiar with the basic operation of the specific machine. When one considers the relatively low cost, the graphics capabilities, the uses of the programming functions and the relatively large memory for program and data storage of these calculators, it is obvious that they can truly enhance a course in statistics.

This manual is not intended as a textbook or study guide, and the reader should be familiar with the basic topics of probability and statistics. Exercises are given throughout the manual for the purpose of checking calculator procedures and program entry. The reader should also refer to the *Owner's Manual* that accompanies the calculator when a more detailed explanation of the functions and keys is necessary.

Although the organization and level of presentation is essentially the same as that of the Preliminary Edition, this manual contains applications for the Hewlett-Packard calculators and many new topics including:

- Programs that sort data, find percentiles, construct box plots and find outliers
- Programs that draw probability histograms for the binomial, Poisson and hypergeometric distributions
- Programs that overlay the normal distribution on the binomial, Poisson and hypergeometric distribution histograms
- Procedures to construct histograms for the HP-28S and HP48 calculators
- Programming techniques for the HP-28S and HP48 calculators
- Simulation and solving applications for the HP-28S and HP48 calculators

It is sometimes quite tedious to type in many of the programs in this manual. The Sharp CE-50P printer/cassette interface and an ordinary cassette tape recorder will allow you to transfer the programs for the Sharp EL-5200 in this manual to student calculators in only a few minutes. Two cassette tapes containing the Sharp calculator programs are available from the publisher to instructors who wish to adopt this manual for classroom use. The contents of the calculator are saved on two cassette tapes since the calculator memory will not hold all of them at once. Tape I includes the programs for descriptive statistics, discrete probability distributions, continuous probability distributions and simulation techniques while tape II contains the programs for sample size, confidence intervals and hypothesis testing. Adopters may redesign the programs in any order on the tapes with the Sharp CE-50P printer/cassette interface and two Sharp EL-5200 calculators by following the instructions in the *Owner's Manual.*

The Hewlett Packard programs in this manual and some of the larger data sets are also available on a computer disk to instructors who wish to adopt the book for classroom use. These may be transferred to the HP48SX or HP48S from either an IBM or Macintosh computer via an appropriate connector cable as described in the *Owner's Manual*. No such option is available for HP-28S users.

This manual is intended for the beginning or experienced Sharp EL-5200 user and for the Hewlett Packard user who is familiar with the basic operation of his or her calculator. Although numerous programs are presented, the sections applying to the HP calculators suggest techniques that can be used in graphing, simulation and solving rather than giving specific programs for all topics in a statistics course.

I hope you will find the programs and techniques instructive and helpful and that they will aid in your understanding and enjoyment of the programmable graphing calculator as applied to the study of probability and statistics.

Iris B. Fetta

June 9, 1991

CONTENTS

CHAPTER 1

INTRODUCTION TO THE SHARP EL-5200

CARING FOR YOUR CALCULATOR

The EL-5200 graphics calculator is fairly easy to use and an amazing tool that you will find quite helpful. Before looking at the various features of this calculator, there are several points you should always remember to ensure that your calculator will work properly. These are:

DO NOT FOLD BACK OR BEND THE COVER CONTAINING THE RIGHT-HAND KEYBOARD.
If you do so, you will probably break the wires connecting the two keyboards. Place the calculator on a flat surface whenever it is open and in use to prevent the cover bending.

DO NOT USE A SHARP OBJECT (SUCH AS A PEN OR PENCIL POINT) TO PRESS THE KEYS ON THE RIGHT-HAND KEYBOARD.
Press the keys with a light touch of your finger or a pencil eraser.

DO NOT DROP YOUR CALCULATOR AND DO NOT CARRY IT AMONG YOUR BOOKS OR IN YOUR PANTS POCKET.
Put it in a protected place if you carry it in your backpack. Do not subject your calculator to a loud noise or severe shock while it is in use - all the keys may become inoperative.

DO NOT EXPOSE YOUR CALCULATOR TO EXCESSIVE HEAT.
For instance, do not leave it on the dashboard of a car in the sun or place it near a heater.

DO NOT USE A CLOTH MOISTENED WITH ANY TYPE OF LIQUID TO LEAN YOUR CALCULATOR.
Keep it clean and free from dust with a soft, **dry** cloth.

GETTING STARTED

The Sharp EL-5200 graphic calculator is a versatile tool and it is not difficult to operate. A liquid crystal display screen and a keypad consisting of plastic dual-function keys are on the left-hand keyboard while the facing right half of the calculator consists of touch-sensitive surface keys. Throughout this manual key-strokes will be enclosed in boxes except for numerical entries and decimal points. These notes are designed primarily for use in a one-semester introductory probability and statistics course. If you are interested in the many other abilities of the Sharp EL-5200, consult your *Owner's Manual*[1].

OPERATION MODES

The EL-5200 has four operating modes that are controlled by the slide switch on the side of the left-hand keyboard. The AER-I (top) and AER-II (second from top) modes are the programming modes, the COMP (third from top) mode is the computation and graphing mode, and the STAT (bottom) mode is the one used for statistical operations. To get started, place the calculator in the COMP mode and press the [ON] key which is located at the top left corner of the left keyboard under the display screen.

DISPLAY SCREENS

The EL-5200 has three display screens: Text (T), Graphics (G) and Data (D). The text screen displays algebraic expressions, answers to calculations and various commands. The graphics screen displays graphs, and the data screen displays statistical data and matrix data.

[1]Sharp EL-5200 Graphic Scientific Calculator *Owner's Manual*, Sharp Electronics Corporation, Mahwah, New Jersey.

You cause the calculator to rotate between these various screen by pressing the [2ndF] and [T-G-D] keys in either the COMP mode or the STAT mode.

❑ OPERATION FUNDAMENTALS

The contrast of the LCD screen display can be adjusted by pressing the (blue) [SHIFT] and [▽] keys, in sequence, several times until the contrast is lowered or the [SHIFT] and [△] keys, in sequence, several times until the contrast is heightened. The contrast should raised enough to clearly see the two small rows of symbols between the LCD screen display and the row of keys containing the [ON] and [OFF] keys.

The keys in these two small rows will darken whenever the corresponding key on the right or left keyboard is pressed. If you press one of these keys by mistake, simply press it again to deactivate the keystroke.

The [2ndF] key is used to activate the second function of the keys that appear in yellow above the keys on the right and left keyboards. There are two [2ndF] keys, one on the left keyboard and one on the right keyboard. Either may be pressed to designate the second function of a key.

The four cursor keys, [▽] , [△], [◁], and [▷], will allow you to move the blinking cursor around the display. To edit something you have typed in, move the blinking cursor to the desired position and keystroke over what is on the display or use the [DEL] and [2ndF] [INS] keys.

If you make an error and an error message appears on the display, press [CL] . You can enter your keystrokes again, but you will probably find it easier to press the [PB] key which appears on the top row of the left-hand keypad. When this key is pressed, the cursor will blink at the position where the calculator found an incorrect keystroke or instruction.

Whenever you are through using your calculator, press the OFF key. If you forget to do so, the calculator will automatically turn itself off when it has not been used for approximately 10 minutes. To resume operation, press the ON key. You will see that the screen contents are, in most cases, exactly the same as when you turned your calculator off.

❒ SETTING THE NUMBER OF DECIMAL PLACES

After clearing the display screen or after the result of a calculation has been displayed, pressing FSE will place the calculator in the FIX (fixed decimal point system), SCI (scientific notation), ENG (engineering notation) or BLANK (floating decimal place) display system. You will probably wish to choose FIX for most of your calculations.

When first entering the COMP mode or after the answer to a calculation or program has appeared on the screen, press TAB followed by the number representing the number of decimal places you wish to use to set the decimal for your computations. The TAB function cannot be activated when in the floating decimal place display system. For now, press FSE until FIX is darkened under the LCD display and press TAB 2. Your calculator display screen should look like

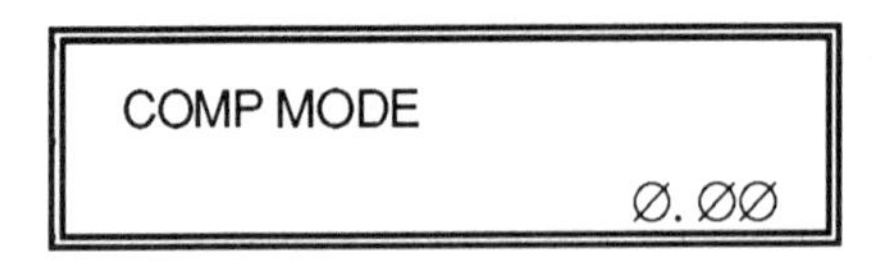

- Note that "zero" appears as Ø to distinguish it from the capital letter "O".

NUMERIC CALCULATIONS

The EL-5200's algebraic operations system allows you to enter numeric expressions as you read them from left to right. If you wish to find the quotient when 3.56 is divided by 4.67, enter 3 . 5 6 [÷] 4 . 6 7 [=] and you will see the result Ø . 7 6 If you desire the result of this calculation to three decimal places, press [TAB] 3. The keystroke sequence to find 7^2 is 7 [x^2] [=], and if you wish to find the square root of 64, press [√] 6 4 [=] .

Notice that the four line display allows you to see the entered expression and the result of the computation. Values appearing on the screen will not affect the result of further computations. If you wish, however, you can clear the screen by pressing the (red) [CL] key.

There is no ± key on the EL-5200. The negative key [(-)] is used to indicate the opposite of a number. The key [−] is used for the operation of subtraction. To enter -3 – 8, press [(-)] 3 [−] 8 [=] and see the result -11. Note that there is a visible difference on the screen when these keys are pressed; the negative key appears on the display as a short dash while the subtraction key appears as a longer dash.

Juxtaposition (implied multiplication) is a very nice feature of the EL-5200. If you wish to multiply 9 by the variable x, you would enter 9X. Note that the variable x is entered using the letter X on the right keypad, not the multiplication sign [×] on the left keypad. Again, there is a visible difference in the symbols on the display screen. Press both the X key and [×] and you will see that the multiplication sign is smaller and slightly raised. If you wished to enter the expression 9(X – 5), press 9 [(] X [−] 5 [)] . Note that 9 [×] X or 9 [×]

(X − 5) could also have been entered, but these expressions require more keystrokes and we do not normally write the products in these forms. The Sharp EL-5200 only understands left juxtaposition. Try entering X 9 or (X − 5) 9. Even though these expressions are equivalent mathematically to 9X and 9(X − 5), the calculator will not accept this syntax.

- If you press = after entering these expressions, you will obtain a numeric value. This is because the Sharp EL-5200 will evaluate an upper-case variable using the value that is stored in that memory location. For instance, if you wished to evaluate the expression 9(X − 5) for X = 8, press 8 STO X 9 (X − 5) = to obtain the value 27.

You should always use parentheses to enclose expressions that need to be evaluated before other computations are performed. If you wish to evaluate $\sqrt{(X + 3)}$ for a particular value of X, say X = 6, to obtain the result 3, you must enter √ (X + 3) . If you enter √ X + 3, you will be evaluating $\sqrt{X} + 3$ for X = 6 and obtain the result 5.449. The use of parentheses will be especially important in entry of programs.

CHAPTER 2

DESCRIPTIVE STATISTICS

■ INTRODUCTION TO GRAPHING ON THE EL-5200

The graphing screen on the Sharp EL-5200 can be accessed in either the COMP mode or the STAT mode. Place the calculator in the COMP mode and for purposes of illustration, press TAB 2 to set two decimal places. Press 2ndF T-G-D to see the graphics screen. This screen is composed of 3072 dots or pixels as indicated in the following diagram:

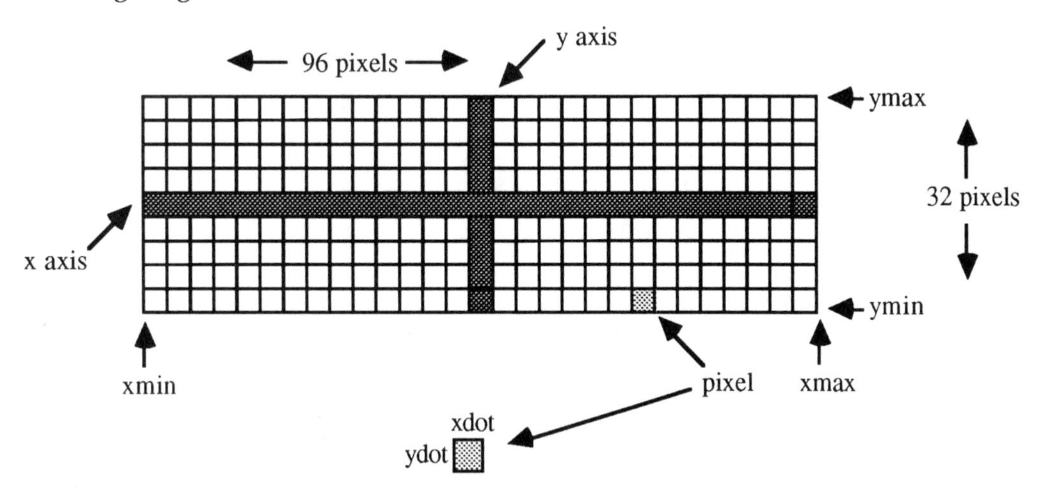

A graph is drawn by the EL-5200 by substituting the x value of each column of pixels into the desired function and displaying the pixel in that column with the calculated y value. This calculator is designed to display additional pixels to provide a visual continuity to the graph.

❐ RANGE PARAMETERS

The range parameters determine the settings for the dimensions of the graphics screen. Press RANGE (located on the top right-hand keyboard of the calculator) and the x-axis range screen is displayed. Press RANGE a second time and the y-axis range screen is displayed. (Pressing RANGE a third time returns you to the text screen.)

Return to the x-axis range screen. Press 2ndF CA to set the "default" range parameters. Use the cursor arrows to move through these settings.

```
X Range
Xmin =
          - 4.7
Xmax =
          4.8
Xscl =
          1.
Xdot =
          Ø.1
Y Range
Ymin =
          - 1.5
Ymax =
          1.6
Yscl =
          Ø.5
Ydot =
          Ø.1
```

Press the RANGE key to exit the range parameters and key in 2ndF T-G-D until you see the graphics screen. Visually verify that the x-axis tic marks are set every 1 unit (you should see 4 of them on the positive x-axis and 4 of them on the negative x-axis since $-4.7 \leq x \leq 4.8$) and that the y-axis tic marks are set every half-unit (you should see three of them on the positive y-axis and three of them on the negative y-axis) since $-1.5 \leq y \leq 1.6$.

Use the RANGE key to return the blinking cursor to the Xmin setting in the range parameters. Enter (-) 10 and press = to store this new setting. Next enter 20 and press = to set Xmax to 20. The blinking cursor should now be over Xscl. Recall that this parameter gives the distance between the marks on the x-axis that will be displayed on the graphics screen. Set Xscl to 2 by entering 2 and =. Press RANGE twice and 2ndF T-G-D to return to the graphics screen. Notice that there are 10 marks on the positive x-axis and 5 marks on the negative x-axis (not counting the origin). Since Xmin = -10 and Xmax = 20, the distance between each mark on the x-axis is 30/15 = 2 = Xscl. Thus, the marks on the x-axis correspond to -10, -8, -6, ... , 16, 18, and 20.

Press RANGE and ▽ ▽ to return to the Xscl parameter. Enter 5 and = to reset Xscl to 5. How many marks should now appear on the x-axis and to which numbers do the marks now correspond? Return to the graphics screen (by pressing RANGE RANGE 2ndF T-G-D) to check your answer.

Press RANGE and press ▽ until you reach the Xdot parameter. (You will probably have the value Xdot = 3.157894737E - Ø1 displayed.) This parameter measures the horizontal width of a single dot or pixel. The values for Xdot can be reset, but when the values of Xmin and Xmax have been input, the value of Xdot is automatically determined by the formula Xdot = (Xmax – Xmin)/95. Thus, resetting Xdot at this point will change Xmax.

- You should try resetting Xdot and Ydot to different values and check the change in the Xmax and Ymax values. Also, return to the graphics screen to see the effect of the changes in pixel width and height. It is recommended that you let the calculator automatically set the values of Xdot and Ydot for the best viewing screen based on the other parameters.

Press [▽] or [RANGE] to access the y-axis range parameters. The definitions of these are analogous to the x-axis range parameters. Set Ymin = -10 (remember to use the [(-)] key, not the [−] key for the negative of a number), Ymax = 10, and Yscl = 5 by entering each value at the blinking cursor position followed by [=]. Note that the formula Ydot = (Ymax − Ymin)/31 holds. (Which numerical values do the marks on the y-axis represent?) Press [RANGE] and [2ndF] [T-G-D] to return to the graphics screen.

❒ TRACING FUNCTION

Enter [PLOT] 10 [,] 5 [DRAW]. There should be a blinking dot on the screen. Press [2ndF] [X⇔Y] (the [△] key) and your screen will look like

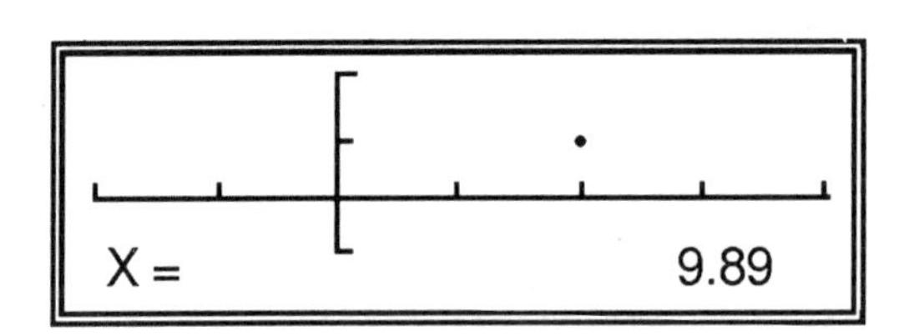

If Xdot had been set to .1 or .2, etc., you would have the cursor blinking directly on x = 10. It is very important to realize that the values displayed using this tracing function (blinking dot) are dependent on the settings of the range parameters and that there may be an error, due to those settings, involved in the actual location that is displayed. Press, and continue to hold down, [◁]. Notice that the blinking dot moves to the left and the x-location of the moving dot is shown at the bottom of the screen. Get as close to the y-axis as possible. If you overshoot and pass the y-axis (x = 0), press [▷] to move the dot to the right and obtain the smallest possible positive x value you can obtain with the current settings of the range parameters. Press [2ndF] [X⇔Y] to see the y-coordinate of this point which should be y = 4.84. (Again, if Ydot had been set to .1, .2, etc., the value of y = 5 could have exactly been determined.) Press and hold [◁] until you can no longer move the blinking dot. Notice that the y value that is displayed does not change since the dot is moving in a horizontal line. Press [2ndF] [X⇔Y] twice to display the x value. You should have reached the minimum value of x, X = -10. By using [▽], [△], [◁], and/or [▷] and the keys [2ndF] [X⇔Y], you can verify the Xmin, Xmax, Ymin, and Ymax values. Press [2ndF] [X⇔Y] until the displayed x or y value at the bottom of the screen disappears to release this tracing function.

Press [2ndF] [G.CL] to clear the contents of the graphics screen. Let's draw the line connecting the points (-5, 5) and (5, 5). Enter [2ndF] [LINE] [(-)] 5 [,] 5 [,] 5 [,] 5 [DRAW] to obtain

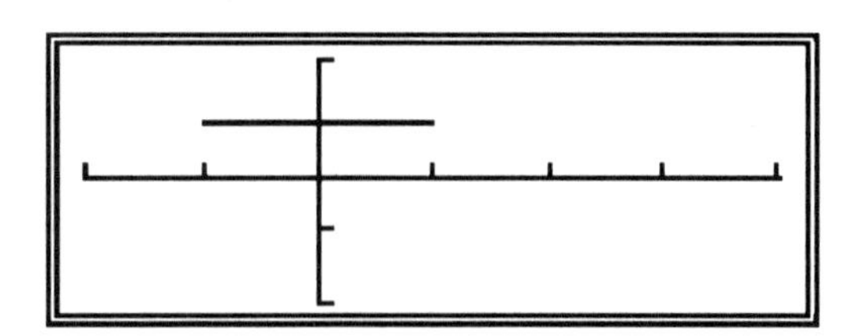

Using [▽], visually move the blinking dot to the point (5, -5). (Recall that the y axis tic marks are set every 5 units.) Press [DRAW]. Using [◁], move the blinking dot to (-5, 5) and press [DRAW]. Complete the square by moving the blinking dot with the cursor keys and pressing draw after you reach each desired location.

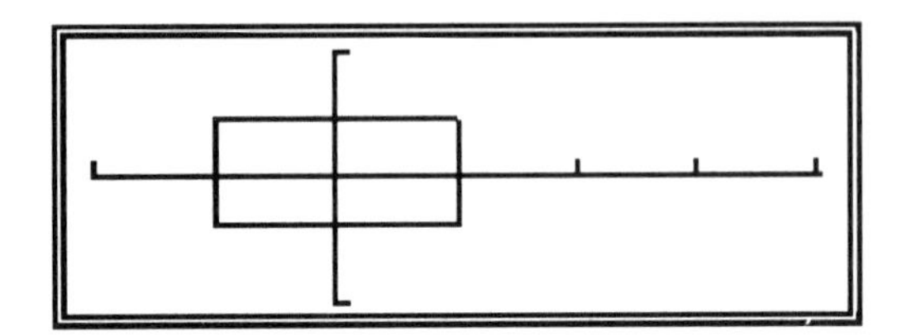

Does your figure look like a square? Using the RANGE key, reset the x-range parameters to Xmin = -30 and Xmax = 30. Exit the range and redraw the square as instructed above to obtain

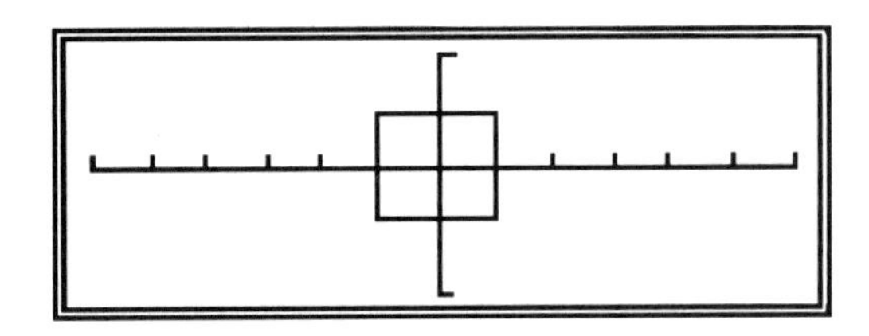

What must be true about the range parameters for the graph of a square to look like a square? (Recall that the screen is composed of 96 pixels in width and 32 pixels in height.)

❑ GRAPHING USING MANUAL RANGE SETTINGS

To graph an equation y = f(x) using manual range settings, set the x and y ranges as explained previously. Press GRAPH , the expression in x to be graphed, and DRAW .

- Note: When entering a function to be graphed, the "y =" portion is not typed in and x must be used as the independent variable.

As an example, let's draw in the lines that are the diagonals (y = x and y = -x) of the previously drawn square. Enter GRAPH X DRAW and GRAPH (-) X DRAW to obtain

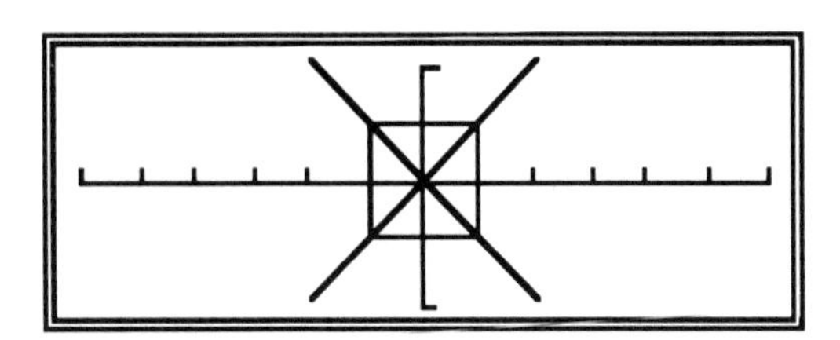

Did you receive an error message? Be sure to use the X key on the right-hand keyboard and *not* the times sign [X]. Also, be certain that you have used the negative symbol [(-)] and *not* the subtraction sign [−] . Press [PB] to locate the source of your error and correct it with the proper keystrokes.

Notice that the calculator can draw two or more graphs on the same graphics screen. When you press [DRAW] , the screen will display "COMPUTING" and the left and right arrow symbols will begin flashing on the bottom line of the display. The amount of time that it takes for the graph to appear depends on the complexity of the function being graphed. The Sharp EL-5200 displays the entire graph at once rather than showing it being drawn point by point.

❐ GRAPHING USING AUTOMATIC RANGE SETTINGS

To graph an equation y = f(x) with the calculator automatically setting the range, enter [GRAPH] expression f(x) [AUTO] Xmin value [,] Xmax value and [DRAW] . As an example, to graph y = x for $-10 \le x \le 10$, enter [GRAPH] X [AUTO] [(-)] 10 [,] 10 [DRAW] . Your screen should appear as

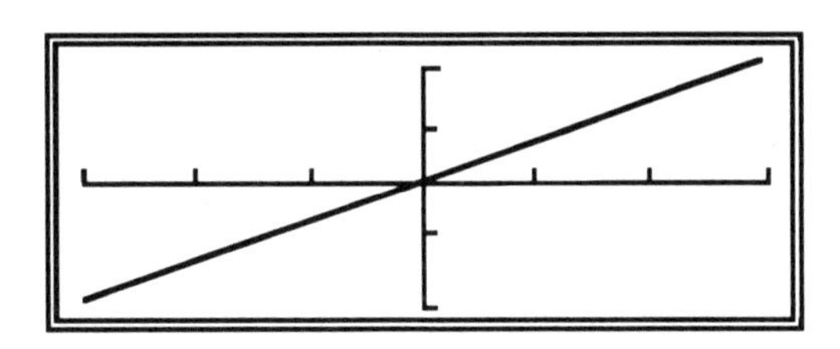

Notice that the graphing scale is different than when you previously drew the line through the square. To see what Xscl and Yscl have been automatically set, press [RANGE] and scroll through the values. Recall that pressing [RANGE] again will cause you to leave the range settings and return to the DATA screen. Now draw in the line $y = -x$ for $-10 \le x \le 10$ using the automatic range settings.

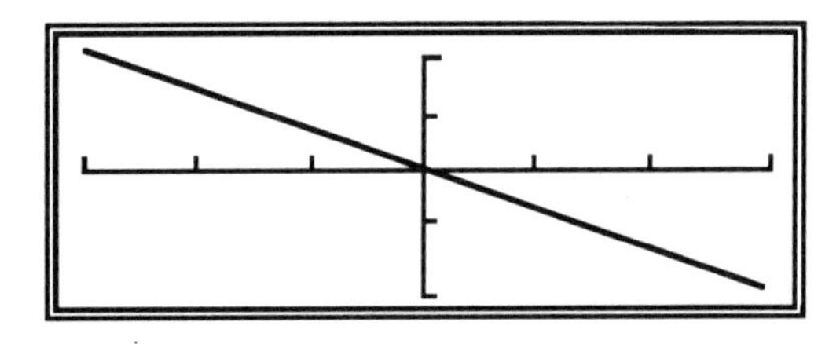

Notice that the graph of $y = x$ is no longer there. If you wish to have both graphs shown, leave off the " [AUTO] xmin value [,] xmax value" and just enter [GRAPH] [(-)] X [DRAW] .

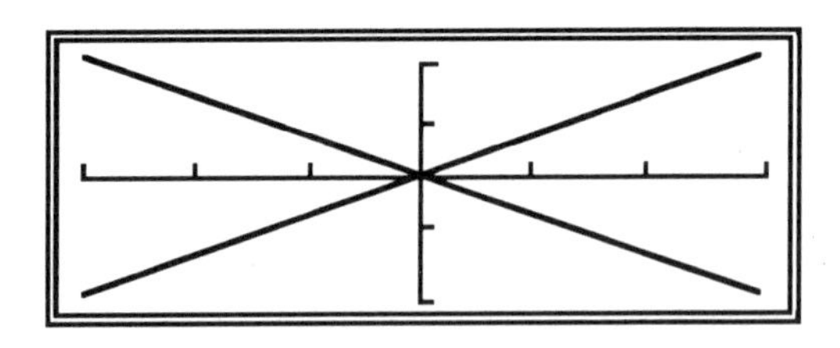

- You do not need to press [2ndF] [CA] to clear the graphics screen since the [AUTO] key resets the range parameters. Whenever the range parameters are reset by you or the calculator, the graphics screen is automatically cleared. If you wish the previously drawn graph to remain on the screen, do not press [AUTO] in the keystroke sequence to draw the next graph(s) and do not reset any range parameters.

EXERCISES

1. Graph $y = (x + 2)^2 - 1$ for $-6 \le x \le 2$ using the automatic range settings.
2. Graph $y = (x + 2)^2 - 1$ for $-6 \le x \le 2$ using the following manual range settings: Xmin = - 6, Xscl = 2, Xdot = .1, Ymin = - 3, Yscl = 2, and Ydot = .5. Use the tracing function to verify that the parabola intersects the y-axis at the point (0, 3). What are the x-intercepts?

HISTOGRAMS

Set the mode selector switch on the left hand side of your calculator to the STAT mode position. The STAT mode contains two submodes: the data store mode and the non-store mode. Press the numeric 1 key to select the data store mode and the numeric 2 key to select the non-store mode.

In the data store mode input data is retained in memory S and results of your statistical calculations are stored in memory Z. When another mode is selected and when the calculator is turned off, the results will remain stored in memory Z. In the non-store mode the input data is *not* retained in the memory.

• Note: It is suggested that the data store mode always be chosen when working in the STAT mode. If your calculator automatically turns off due to nonuse for approximately 10 minutes or if you switch to another mode, you will not lose the data you have input in the STAT mode.

Statistical graphs are drawn with the calculator in the STAT mode and in the data store mode all data entered for a statistical calculation are stored in memory S. Array memory S is write-protected in that new data values are not automatically written over values already stored in S. Thus, whenever memory array S is used, it must be cleared manually or an error message ("error 5") will result.

Place the calculator in the STAT mode and press 1 for data store. If you see that there are data values stored in the S array, that is, you will have something other than "S: [0,0], P" on the screen, follow the procedure described next to delete this data from the calculator.

❐ CLEARING THE S MEMORY LOCATION

The following procedure will clear the S memory location:

- Select the STAT mode. Press [TITLE / DATA] to select the data screen and if necessary, use [△] and/or [▽] to position the blinking cursor over the letter S in the S array.

- Press [2ndF] [CA] and press [ENT] when the message "CLEAR ? → ENT" appears.

- Press [2ndF] [T-G-D] once to return to the text screen.

- Note: Statistical memory S must be cleared *each* time the histogram procedure is used.

❐ ENTERING SINGLE-VARIABLE DATA

To enter single variable data, press [2ndF] and [T-G-D] to place the calculator in the text screen. To enter data values one by one, press (white) [DATA] (the [M+] key) after each numerical entry. To enter two or more of the same numerical values, enter the values separately or press [X] followed by the numeric key representing the frequency of the data value. Note that the cumulative count of the number of data values appears on the right hand side of the screen after each entry.

Consider the following data representing the amounts of regular unleaded gasoline (rounded to the nearest gallon) purchased by 20 customers at a local service station:

15	14	19	10	17
10	20	18	6	23
8	11	13	10	14
9	17	12	14	12

To enter the first value, press 1 5 [DATA] 1 4 [DATA] 1 9 [DATA] , etc. Notice that you do *not* press the second function key before pressing the data key since DATA is not written in gold letters. Enter the rest of this data into your calculator following this same procedure. When you finish entering the data, your last screen should be similar to

```
14 DATA
                19.ØØ
12 DATA
                20.ØØ
```

Recall that the values on the left of the screen are the data values that you have entered and the values on the right indicate the number of values that you have entered. It is always a good idea to check that the last value on the right equals the total number of data values that are to be entered (in this case, 20).

CHECKING THE ENTRY OF DATA

To verify your data, access memory S by pressing TITLE [DATA] and [=]. Scroll through the entries by pressing [△]. The data values will appear on the right hand side of the screen.

NAMING DATA

To name your data set after the data has been entered, press TITLE [DATA] followed by [▷]. Enter your title using the alphabetic keys on the right hand side of the keyboard and press [=] .

DISPLAYING VALUES IN THE STORE MEMORIES

If you wish to see the values in the S memory array, press TITLE [DATA] and position the blinking cursor over the S. Press [=] SET and the data values will appear on the right-hand side of the display in the reverse order in which they were entered. Press and hold [△] to scroll through the values. Press TITLE [DATA] and return to the text screen with [2ndF] [T-G-D] .

DATA ENTRY CORRECTION

If a mistake is realized before pressing [DATA] , press [◁] or [▷] to move the cursor over the incorrect entry and enter the correct value. If a mistake is realized after the [DATA] key has been pressed, call up the data as instructed in "checking the entry of data" and move the cursor with the [△] and [▽] keys until the cursor is blinking on the position of the data entry to be corrected. Enter the correct value and press the [=] key.

❑ CONSTRUCTING THE HISTOGRAM

When in the STAT mode, there is a parameter called Δx in the range settings that is not available in the COMP mode. This setting is used to determine the (equal) width of the class intervals (the bases of the rectangles) of the histogram.

$$\Delta x = \frac{X_{max} - X_{min}}{\text{the number of classes}} = \text{width of each class}$$

To have the calculator automatically construct a histogram of data values, press [2ndF] [G(HI)] [G(HI)] [AUTO], and [DRAW]. The resulting histogram probably will not be what you would consider a "good-looking" histogram. Press [RANGE] and notice that Xmin = 6 (the smallest value in the gasoline data), Xmax = 23 (the largest value in the gasoline data), and Δx = 1.717894737. (Solve the above formula for the number of classes using these values for Xmin, Xmax and Δx and you will find that the calculator is showing 9.9 intervals. It is difficult to visually discern 9.9 intervals from 10 intervals.)

- When using the automatic range settings in the Sharp EL-5200, the class width Δx is determined by the formula (Xmax − Xmin + Xdot) ÷ 10. This setting allows the value for Xmin to be the smallest data value and the setting for Xmax to be the largest data value. Usually the value of Xdot is very small and will not be used in the formula for the class width in this manual.

- It is not necessary to set the parameter Xscl to a certain value when drawing histograms. The interval widths are determined by Δx. If you wish to try to visually count the frequencies of the data (the number of measurements falling in each interval), you may set Yscl to an appropriate value.

Also note that the automatic range settings give Ymax = 31 unless the frequency in any particular class exceeds 31. If so, the Ymax value is set to that highest frequency. For most histograms you construct, you will need to reset the range parameters to zoom in on the graph for a better view. To reset the automatic settings, press RANGE and use the up and down arrow keys to scroll through the settings. With the cursor on the setting to be changed, enter the new setting value and press =. Reset Ymax to 5 and Yscl = 1 for this example and return to the screen you were on before entering the range settings by pressing RANGE . Redraw the histogram with the keystroke sequence 2ndF G(HI) DRAW . (After manually resetting the range parameters, do not press AUTO when drawing the histogram or you will undo what you have reset in the range.) The resulting histogram looks much better, doesn't it?

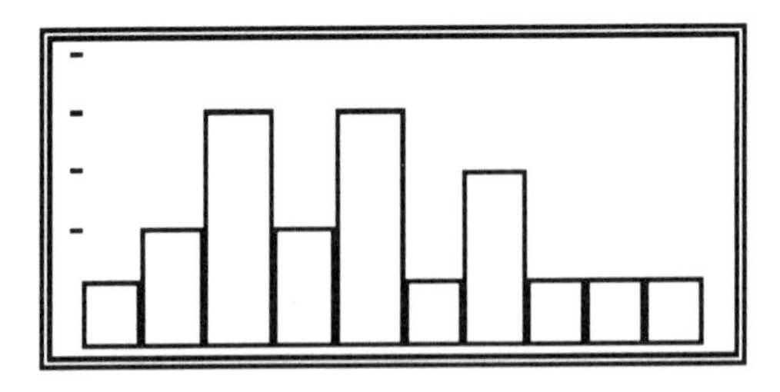

❑ DETERMINING THE CLASS LIMITS AND CLASS FREQUENCIES

The tracing function gives the frequencies (heights = number of data values in each class) of the class intervals, from left to right, when the [▷] key is depressed repeatedly with a histogram drawn on the graphics screen. Depression of the [2ndF] [X⇔Y] keys twice gives the x value of the *left* endpoint of the interval (lower class limit).

Suppose you wished to construct a histogram consisting of 5 intervals for the gasoline data. Calculate, using the formula for Δx, the class interval width. Did you get $\Delta x = 3.4$? Set this value for Δx in the range and redraw the histogram with the keystrokes [2ndF] [G(HI)] [DRAW] .

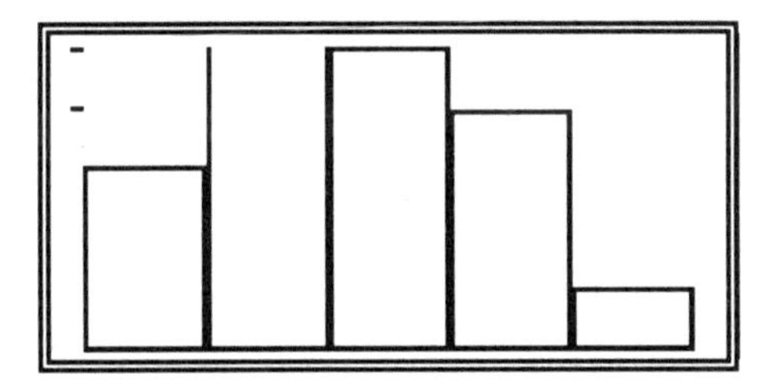

Notice that the top of the second rectangle does not show in the screen view. Recall that Ymax is set at 5. Use the tracing function ([▷] key) to determine the frequency of the second class. Did you notice the screen scroll upward so that the top of the second rectangle comes into view?

Press [2ndF] [X⇔Y] to release the tracing function. Notice that the bottoms of the rectangles ar no longer visible. To obtain a "good-looking" histogram, press [RANGE] twice and reset Ymin to Ø and Ymax to 7. (Setting Ymax to 6 would also give a nice graph.) Exit the range and redraw the histogram with [2ndF] [G(HI)] [DRAW] .

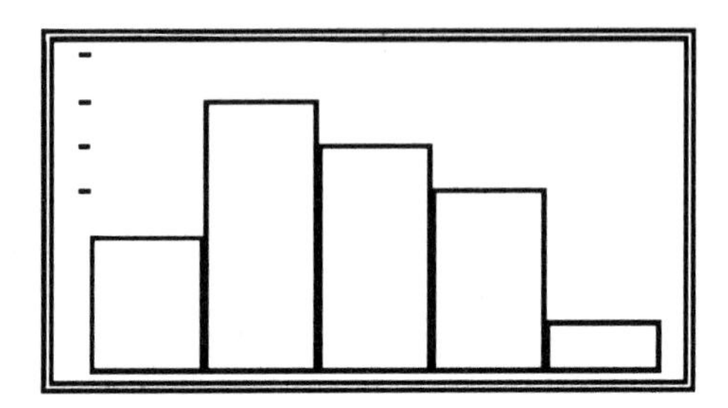

To determine the class limits and frequencies for this histogram, press [▷] repeatedly and record the y values (frequencies). Press [◁] until you return to the first frequency of 3. Press [2ndF] [X⇔Y] twice and use [▷] to display the lower class boundaries. (Remember that Xmin = 6 is the lower boundary of the first class.)

- If you look closely at the screen, you will see a blinking dot in the upper right-hand corner of the rectangle for which the frequency is displayed. When you are displaying the class limits (x values), the blinking dot will be in the upper *right*-hand corner of the rectangle for which the *lower* class limit is being displayed. Since there are no gaps between the rectangles, the lower limit of each class, except the first, will be the upper limit of the preceding class.

The resulting frequency distribution is:

GALLONS OF GASOLINE PURCHASED	FREQUENCY
6.0 - 9.4	3
9.4 - 12.8	6
12.8 - 16.2	5
16.2 - 19.6	4
19.6 - 23.0	1

Unfortunately, there is a problem. There were originally 20 data values and the sum of the frequencies in the above distribution is only 19. Why? The EL-5200 will place a data value that falls on a boundary in the next class. Thus, even though you did not see the sixth rectangle on your screen, it was there and it contained the data value 23. How can you solve this problem? Depending on what you feel gives the "best-looking" histogram, you can

- Round Δx *up* so that the width of each class increases and no data value falls on a boundary. Xmax should be reset to Xmin + k(Δx) where k is the number of classes used so that the last rectangle will be completely visible on the screen.

- Reset Xmin to a value slightly smaller than the minimum data value. If Xmin is reset to a value that is kept to one more decimal place than the data values and Δx is such that Xmin + Δx does *not* sum to a possible data value, you will never have a data value falling on a class boundary. Again, Xmax should be reset to Xmin + k(Δx) where k is the number of classes used so that the last rectangle will be completely visible on the screen.

For example, let's redraw the histogram for the gasoline data using the settings Xmin = 5.5, Δx = 4, Xmax = 5.5 + 5(4) = 25.5. You should already have the settings Ymin = 0, Ymax = 7, and Yscl = 1.

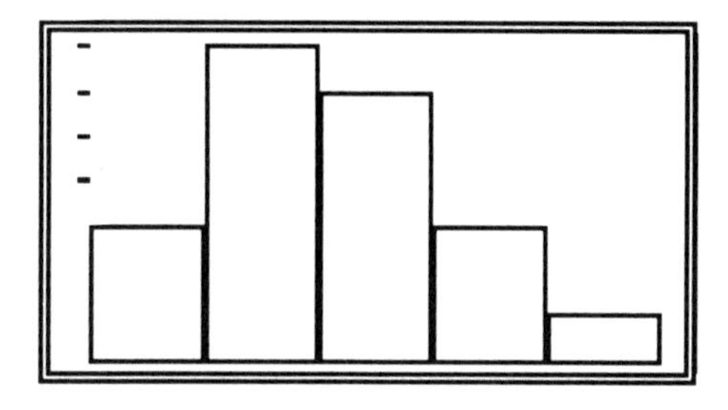

The resulting frequency distribution is:

GALLONS OF GASOLINE PURCHASED	FREQUENCY
5.5 - 9.5	3
9.5 - 13.5	7
13.5 - 17.5	6
17.5 - 21.5	3
21.5 - 25.5	1

• Note: The number of class intervals is automatically set to 10 when the AUTO key is used in drawing a histogram. The maximum number of class intervals permissible for a histogram drawn on the Sharp EL-5200 is 32.

This data will be used again in the next section of the manual. To save reentering it, transfer the data to memory A using the following procedure so that it can be recalled at a later point.

❐ TRANSFER OF STATISTICAL DATA

If you wish to keep the data that is stored in memory location S, it may be transferred to another unused memory location (for example, array A). Set the mode selector switch to COMP. Press 2ndF MATRIX (b) to place the calculator in the MATRIX mode. Press MAT S STO MAT A and press TITLE/DATA .

To reverse the procedure and place the data back in memory S for use in the STAT mode, follow the same procedure as above replacing the letter S with the letter A. If you obtain an "error 5" message, you forgot to clear array S before trying to transfer the data.

EXERCISES

The annual amounts of rainfall for 30 randomly selected cities are recorded below (in inches):

12.0	9.4	15.4	12.3	10.6
13.6	11.7	9.8	11.0	10.1
17.6	8.1	13.2	8.8	12.1
10.9	8.9	13.1	11.2	10.4
12.2	10.9	9.7	10.3	9.2
10.2	12.5	10.5	14.5	10.5

1. a) Construct a histogram for this data. Use a class interval width equal to 1.2.
   b) How many rectangles compose your histogram?
   c) Use the tracing function to give the heights of the rectangles, in order from left to right.
   d) What skewness is shown?

2. a) Set the class width equal to 5 and draw the histogram.
   b) What effect does this new setting for Δx have on your histogram? Explain in terms of the calculation formula for Δx. Look closely at the last interval width shown on your calculator screen. Does what you visually see agree with the number of intervals you found using the formula for the class width?
   c) Use the tracing function to find the frequency of the each of the class intervals.
   d) What skewness is shown?

3. a) Set the class width equal to 2 and draw the histogram. What effect does this new setting for Δx have on your histogram?
   b) Explain in terms of the calculation formula for Δx. Look closely at the last interval width shown on your calculator screen. Does what you visually see agree with the number of intervals you found using the formula for the class width?
   c) Use the tracing function to find the frequency of each of the intervals.
   d) What skewness is shown?

4. Determine the settings for the range parameters that will give what you consider the "best-looking" histogram for this rainfall data.

BROKEN-LINE GRAPHS

A broken-line graph or frequency polygon is a line graph formed by connecting, with straight line segments, the points obtained by taking a horizontal value within each of the class intervals and the corresponding vertical value equal to the respective class frequency.

Select the STAT mode and clear memory array S by pressing [TITLE/DATA] [2ndF] [CL] [ENT] . Set the mode selector switch to COMP. Recall the gasoline data from the previous section with the following keystroke sequence: [2ndF] [MATRIX (b)] [MAT] A [STO] [MAT] S [TITLE/DATA] . Return to the STAT mode and redraw the histogram with the automatic range settings, enter the range, set Ymax to 5, and draw the histogram again with [2ndF] [G(HI)] [DRAW]. With the graph of the histogram on the screen, press [2ndF] [G(BL)] [DRAW] to obtain the line graph. The broken-line graph for the gasoline data when superimposed on the graph of the histogram is

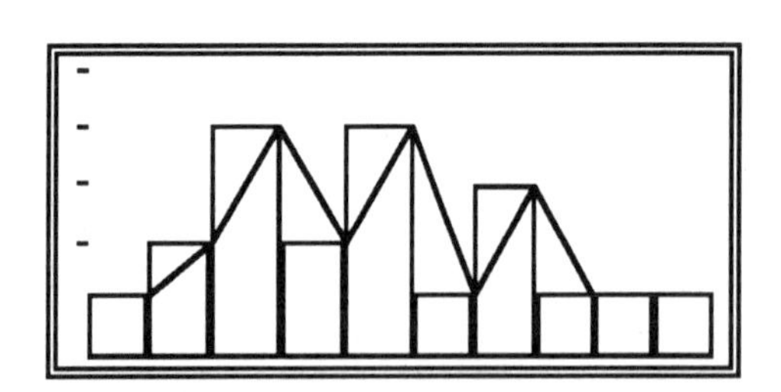

Broken-line graphs may also be drawn without viewing the histogram by pressing [2ndF] [G(BL)] [AUTO] [DRAW] . Recall that as with the histogram, you will have to probably have to reset the value of Ymax in the range to obtain a visually

pleasing graph. You could also construct the histogram, press [2ndF] [G.CL] to clear the graphics screen, and then key in [2ndF] [G(BL)] [DRAW] to obtain the broken-line graph without the histogram being shown.

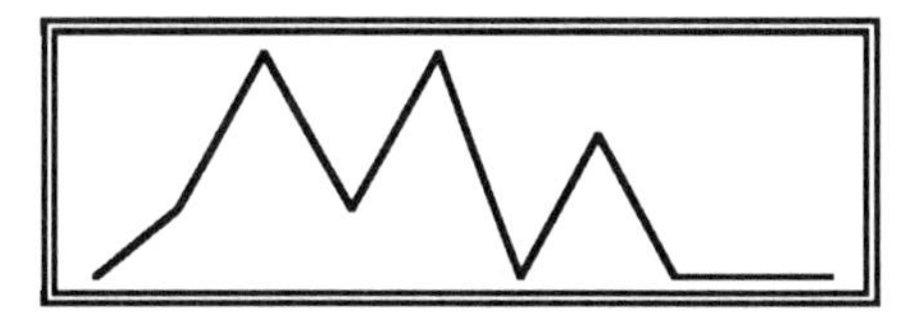

The standard convention is to take the midpoint of each class as the (horizontal) x value. The Sharp EL-5200, however, chooses as the x-coordinate as the value at the left endpoint of each of the rectangles (class intervals). Recall that the blinking dot on the screen shows this value at the right endpoint of each of the rectangles. Construct the histogram for the gasoline data using the range parameters Xmin = 5.5, Δx = 4, Xmax = 25.5, Ymin = 0, Ymax = 7, and Yscl = 1 and overlay the broken-line graph to obtain

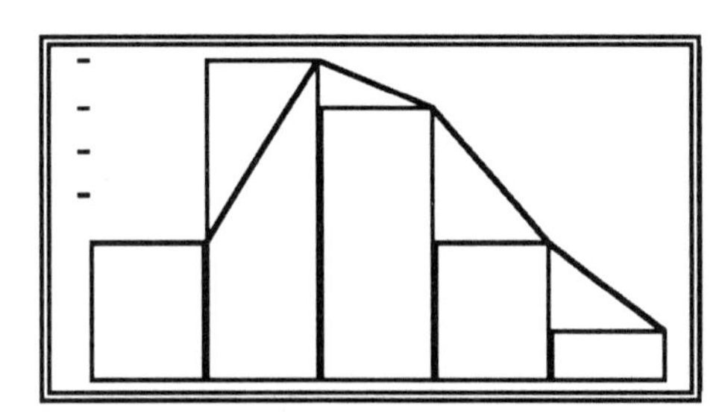

If you use the tracing function to find the coordinates of the endpoints of the line segments, you will find the leftmost endpoint as (5.5, 3). Recall that 5.5 is the lower limit of the first class interval. Confusing, but the general shape of the broken-line graph is roughly the same as if constructed by hand. Drawn without the histogram, the graph appears as

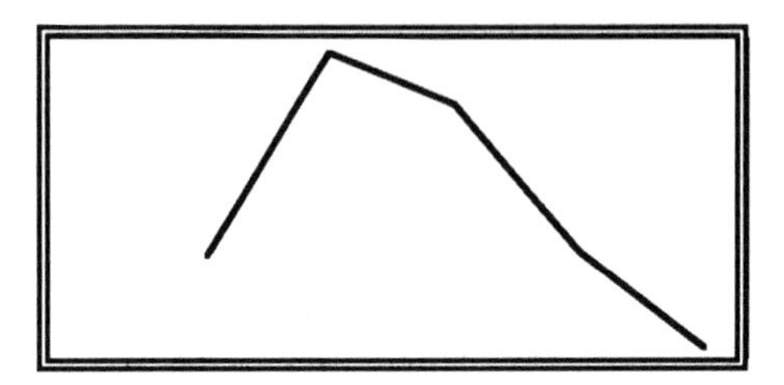

Overlaying histograms for two different distributions on the same graph can be confusing. When comparisons are needed, broken-line graphs are usually preferable.

CUMULATIVE FREQUENCY GRAPHS

A cumulative frequency distribution is derived from a frequency distribution and provides another graphical description of qualitative (numerical) data. The cumulative frequency graph is constructed in the same manner as the broken-line graph except that the y-coordinate of each segment endpoint represents the cumulative (accumulated) frequency of measurements less than or equal to the upper boundary of each class. A cumulative frequency distribution for the gasoline data of the previous section (derived from the frequency distribution on page 25) is

GALLONS OF GASOLINE PURCHASED	CUMULATIVE FREQUENCY
5.5 - 9.5	3
9.5 - 13.5	10
13.5 - 17.5	16
17.5 - 21.5	19
21.5 - 25.5	20

The cumulative frequency graph is drawn by pressing [2ndF] [G.CF] [DRAW] and appears as an overlay if the histogram is already on the screen. Note that you will have to reset Ymax to 20 to see the entire cumulative frequency graph.

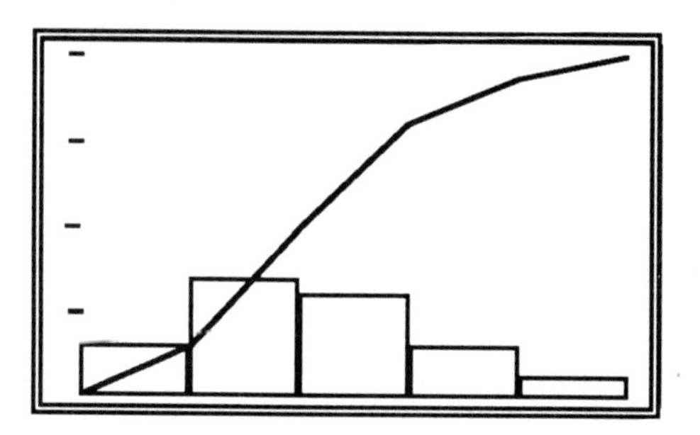

Cumulative frequency graphs can be drawn alone, over histograms, or over broken-line graphs. To draw a cumulative frequency graph without viewing the histogram, press [2ndF] [G.CL] [2ndF] [G(CF)] [AUTO] [DRAW]. You could also construct the histogram, press reset the Ymax value (which will clear the graphics screen), and then key in [2ndF] [G(CF)] [DRAW] to obtain the cumulative frequency graph without the histogram being shown.

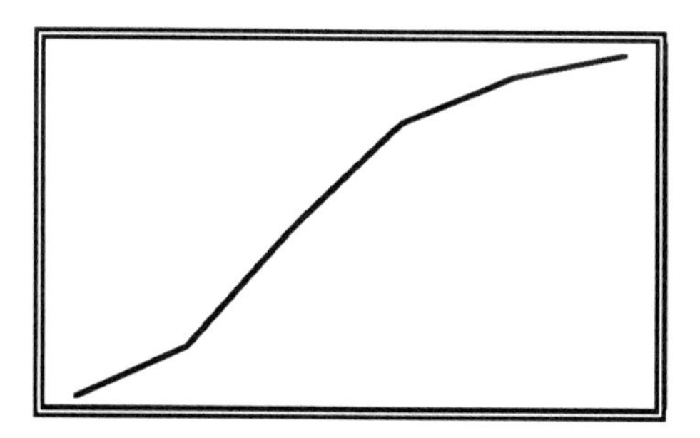

Experiment with different combinations of graphs.

NUMERICAL DESCRIPTIVE MEASURES

The Sharp EL-5200 must be in the STAT mode in order to use the single-variable statistics function keys. (Two-variable statistics functions will be discussed in Chapter 9.) The single-variable statistics function keys, the keystrokes required to obtain those functions, and what the functions return as information are:

- [n] (press [2ndF] Ø) the number of entered data values
- [Σx] (press [2ndF] .) the sum of the entered data values
- [Σx^2] (press [2ndF] [(-)]) the sum of the squares of the entered data values
- [$\bar{x}$] (press [2ndF] 4) the mean of the entered data values when those values constitute a sample
- [Sx] (press [2ndF] 5) the standard deviation of the entered data values when those values constitute a sample
- [σx] (press [2ndF] 6) the standard deviation of the entered data values when those values constitute a population

- Note that [σx] [x^2] will give you the variance of the entered data values when those values constitute a population.

- Also note that [$\bar{x}$] will give the population mean μ if the entered data represents a population.

Basic calculations may still be performed in the STAT mode. However, the independent accessible memory (the RM , ⇒M and M+ keys) *cannot* be used since the functions labeled in white above these keys, CD , (x, y) , DATA , are activated while in the STAT mode.

Recall that to enter single-variable statistical data, you first must enter the STAT mode. Choose 1 for data store or 2 for non-store.

- In the data store mode, all statistical data you enter will be saved in memory array S until it is cleared manually. In the non-store mode, the input data itself is *not* retained in memory and will be lost if the calculator is placed in another mode, if it is turned off or if it is in the AUTO POWER OFF state due to nonuse.

To enter data values one by one, press (white) DATA (the M+ key) after each numerical entry. To enter two or more of the same numerical values, enter the values separately or press ☒ followed by the numeric key representing the frequency of the data value. If you would prefer, press DATA consecutively to enter repeated values. Note that the cumulative count of the number of data values appears on the right hand side of the screen after each entry.

In both the data store and non-store modes, the results of the statistical calculations are stored in memory array Z. This data will remain regardless of the mode you are in until it is replaced by recalculation of the basic statistics.

Memory Area	Z[1]	Z[2]	Z[3]	Z[4]	Z[5]	Z[6]
Stored Statistic	n	Σx	Σx^2	Σxy	Σy	Σy^2

- If single-variable data is entered, memory locations Z[4], Z[5], and Z[6] are empty. If you wish to see the values stored in memory array Z, return to the COMP mode, press [TITLE / DATA], and use [▽] until the blinking cursor is over Z. Press [=] and these values will be displayed. Press [2ndF] [T-G-D] to return to the text screen.

If you wish to enter additional data for the same problem, simply enter the values as described above and they will be placed at the end of the entered data. If you wish to begin with another data set, statistical array S must first be cleared.

EXERCISES	ANSWERS
1. Find for the gasoline data on page 17, a) the sample size b) the sample mean c) the sample standard deviation d) the sample variance	1. a) 20 b) 13.60 c) 4.382 d) $4.382^2 = 19.202$
2. A small population consists of the following data values: 0, 2, 2, 2, 3, 4, 4, 5, 5, 5, 6, 6. Find a) the size of the population b) the population mean c) the population variance d) the actual percentage of the data values falling in the interval $\mu \pm 2\sigma$ e) the sum of the squares of the data values	2. a) 12 b) $\mu = 3.667$ c) $1.795^2 = 3.222$ d) $\mu \pm 2\sigma =$ (.077, 7.257) which contains all values except 0. Thus, 11/12 = 91.67% of the data is in this interval. e) 200

PROGRAMMING ON THE SHARP EL-5200

As this will probably be your first programming experience using this calculator, let's take a moment to look at the function called the Algebraic Expression Reserve (AER). The AER-I and AER-II modes allow programming that is relatively simple and logical as well as being a convenient tool for performing repetitive calculations. The calculation steps are executed in the COMP mode.

The AER-I mode is useful for writing programs that repeatedly use the same variables, while the AER-II mode allows the use of lower case letters, Greek letters, and other special characters as variables. A maximum of 99 programs can be stored in the EL-5200. The programming capacity is approximately 5K (5120 bytes), and each program has a maximum length of 160 steps (bytes).

Variables may consist of uppercase letters whose values are stored, respectively, in store memories A through Z. One uppercase letter may be multiplied by another by entering the variables next to each other (for example, AB) without having to use the multiplication sign ([×]). You may also use the value stored in the memory for the variable directly in a particular calculation in the COMP mode by simply entering the letter of the variable in the calculation. In the AER-II mode, lowercase letters may be used as variables and may be combined to create names of variables. For this reason, lowercase variables must have the multiplication sign between them to indicate multiplication, and you cannot use a combination of upper and lowercase variables for the name of a single variable.

The assignment of values to variables in the two modes differ. For example, in the AER-I mode, if you wish to let B = 2, you would keystroke 2 [STO] B. In the AER-II mode, to assign the value of 2 to b, you would enter b = 2.

In both the AER-I and AER-II modes the [f()=/?] key is used for prompting of uppercase variables. When this key is first pressed, the " f(" appears on the calculator display. The variable(s) to be prompted are then entered and [f()=/?] is pressed again to show the closing of the parentheses ") " on the display screen. It is not necessary to enter a comma ([,]) or space ([␣]) between several variables in the AER-I mode when using [f()=/?] unless you wish to have them to make the programs more readable. As you will see in the following program, when used in the AER-I mode the [f()=/?] key results in the prompt "variable = ?". (The use of [f()=/?] and the prompting of lowercase variables will be discussed in a later section.)

❐ EDITING PROGRAMS

It is relatively easy to overlook a keystroke or press an incorrect key when typing in a program. Therefore, it is strongly recommended that you look very carefully at the "exactly what your calculator screen should look like" box and test each program with the exercises provided before using any program with your own problems.

❐ ERROR MESSAGES

If you receive an error message while running any program in the COMP mode, press and hold the [PB] key. The blinking cursor will appear at the source of the error. The procedures for correcting the mistake are explained in the following sections. Some common mistakes to avoid are:

- Using the negation sign [(-)] (which appears as a short dash on the display screen) rather than the minus sign [−] (which appears as a longer dash) or vice-versa,

- Using a zero (Ø) instead of the letter O (without a slash) or vice versa,

- Using the capital letter X (large) or the lowercase letter x (script) in place of the times sign [×] or vice versa,

- Typing the letters for a calculator function, such as ENT (which results in the implied multiplication of the three variables, E, N, and T), instead of pressing the [ENT] key.

If you do not receive an error message but you also do not obtain the correct answer to the exercises, you have probably made an error with the parentheses in your program entry. After displaying the program contents on the screen (see below), compare what is on your screen with the printed, boxed form of what your program should "look exactly like" that is given following each of the programs. Be certain to check the symbol at both the beginning and the end of each line. If these do not exactly match, then you have an extra character or you have omitted a character somewhere in that line. Other common mistakes including misreading a division sign for an addition sign and forgetting to store the corrections after they are made by use of the enter key.

- Note that the (white lettering) [ENT] key and the green key [COMP] are exactly the same key. You do not press [2ndF] before pressing this key if you wish the [ENT] key function because ENT is not labeled in gold letters. The instructions in this manual will refer to this key as either [ENT] or [COMP] depending on the particular function desired.

❐ CORRECTING A MISTAKE WHILE ENTERING A PROGRAM

If you realize a mistake while entering a program, simply use the cursor keys [△], [▽], [◁], and/or [▷] to reposition the blinking cursor and key in the correction.

❐ EDITING A STORED PROGRAM

To edit a program that has been previously stored, place the calculator in the mode in which the program was entered. Search for the program you wish to edit by pressing TITLE [PRO] and scroll through the titles by pressing and holding this [PRO] key. When the desired title is displayed, press [▽] which will cause the contents of the main program routine to appear for editing. If the program required more than four lines, you can use the cursor keys to scroll through the program.

❐ INSERTING AND DELETING CHARACTERS

If you need to insert a character, use the cursor key(s) to position the blinking cursor to the immediate right of where you wish to insert a character. Press [2ndF] [INS] and then press the key that corresponds to the character to be inserted. If more than one character is to be inserted, you must press [2ndF] [INS] before each insertion. To delete an entry, position the blinking cursor over the entry to be eliminated and press [DEL] . *Do not forget to press* [ENT] *after your corrections have been completed to store the corrected program.*

❐ EDITING A PROGRAM TITLE

To edit a stored program title, select the mode in which the program was entered and display the title to be edited with the [PRO] key. When the title you wish to edit is displayed, press [▷]. Using the cursor key(s), move the blinking cursor to the position to be changed and type in the corrected title. Press [ENT] to store the edited title.

DELETING A PROGRAM

To delete an entire program from the calculator memory, select the mode in which the program was entered. Using PRO , display the title of the program you wish to delete. Press 2ndF CA and then press ENT as instructed on the screen to delete (clear) the program and program title. (Press CL if you change your mind and don't wish to delete the program.)

BOX PLOTS

Box plots, sometimes called box-and-whisker plots, provide a useful graphical technique for describing data. These graphs use information regarding how the measurements are spread over the interval from the smallest value to the largest value in a set of data.

PERCENTILES

When you describe the position in relation to the other measurements of a particular measurement in a data set, you are using a measure called a *percentile*. When your data set is arranged in order from smallest to largest, the pth percentile is a number (which may or may not be one of the data values) such that p% of the measurements fall at or below that number. The median of a data set is the 50th percentile.

The 25th percentile is called the *lower quartile*, q_1, and is the median of the lower half of the data. The 75th percentile is called the *upper quartile*, q_3, and is the median of the upper half of the data. The *interquartile range*, iqr, is the difference $q_3 - q_1$ and tells us the spread of the middle half of the data.

In order to use your Sharp EL-5200 calculator to find percentiles for a set of data, the measurements must be placed in the calculator in increasing order since the Sharp EL-5200 does not have a built-in sorting routine.

The following program, Sort, prompts for entry of data and sorts the data in ascending order. Experienced Sharp EL-5200 users should go to the "your calculator screen should look exactly like the following" boxes when entering programs.

PROGRAMMING

Select the AER-II mode. You will see on the display Ø1: TITLE? Type in, using the letters on the right keyboard, SHIFT S o r t and press ENT to store the program name. Input the program at the M: prompt as follows:

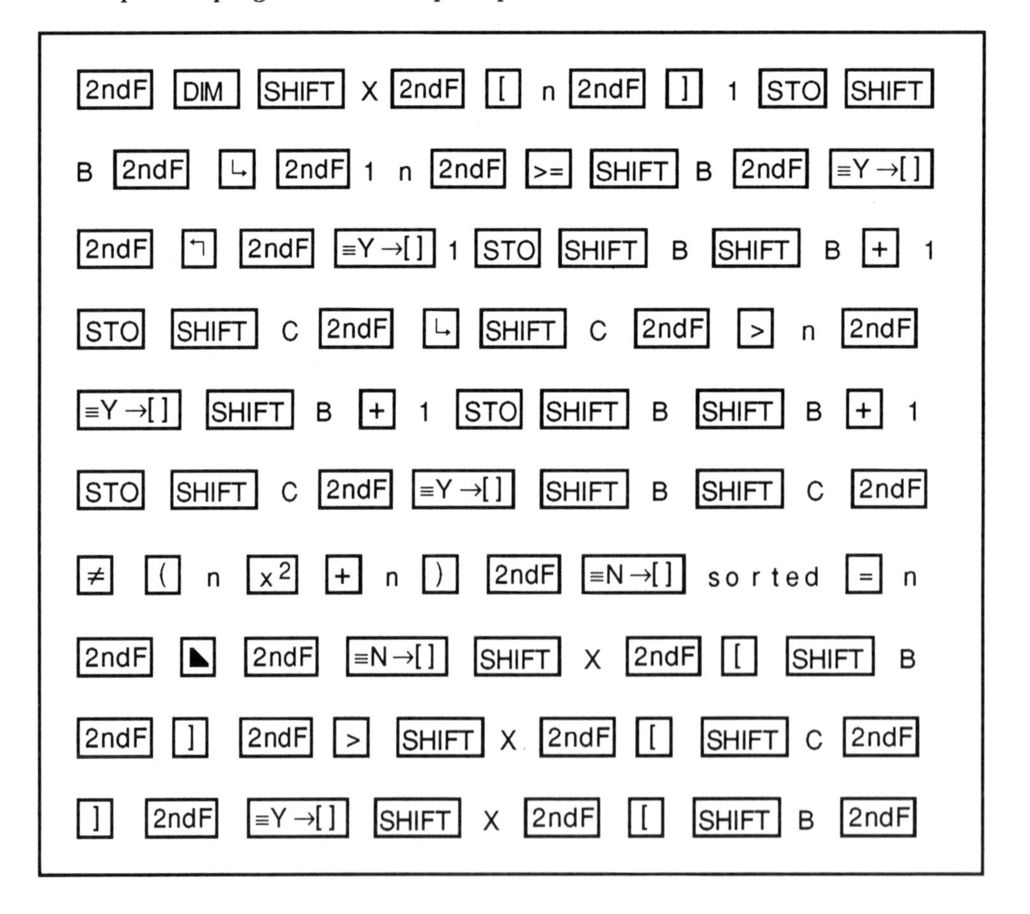

Program Sort is continued on the next page.

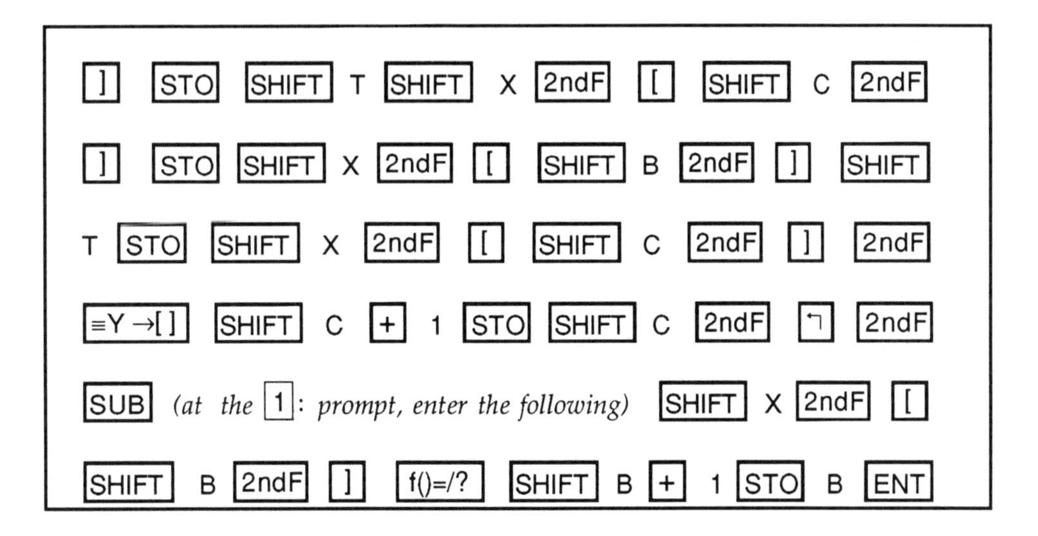

Once the program is entered, your calculator screen should look exactly like the following:

```
M: DIM X[n]1⇒B↳ 1 n
>=B ≡Y→[ ↰ ]1 ⇒B B +1 ⇒
C ↳ C > n ≡Y→[ B +1⇒ B B +
1⇒ C] ⇒ BC ≠ ( n2 + n ) ≡N→
[sorted=n◣] X[B]⇒T X[
C]⇒ X[B]T⇒X [C] ] C+
1 ⇒C ↰
```

Press 2ndF ▽ to see the contents of subroutine 1 which should appear on your screen exactly like this:

```
1: X[B] = ?B+1 ⇒B
```

- Program Sort may take a while to run if the data set is large and requires a lot of sorting. Its use can be avoided if the data is first sorted by hand and entered in ascending order into matrix location X . Program Sort is probably faster and more efficient than if you do sort the data by hand, however.

Program Percentile calculates percentiles for the sorted statistical data.

PROGRAMMING

Select the AER-II mode. You will see on the display Ø2: TITLE? Type in, using the letters on the right keyboard, SHIFT Pe r c e n t i l e and press ENT to store the program name. Input the program at the M: prompt as follows:

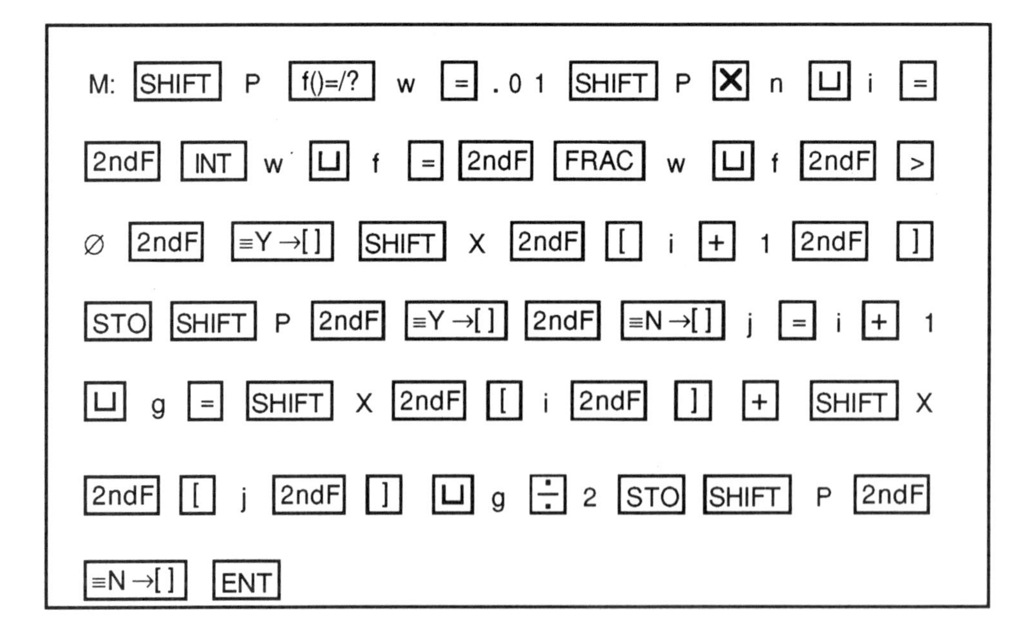

Once the program is entered, your calculator screen should look exactly like the following:

```
M: P = ? w = . 0 1 P x n ⊔ i =
INT w ⊔ f = FRAC w ⊔ f
> Ø ≡Y→[ X [ i + 1 ]⇒P ] ≡
N→[ j= i+1 ⊔ g = X [ i ] +
X[ j ] ⊔ g ÷ 2 ⇒P]
```

EXERCISES	ANSWERS
For the 12 data values 30, 23, 56, 24, 25, 50, 24, 30, 21, 44, 45, 32, find the	
1) 25th percentile	1) 24
2) 80th percentile	2) 45
3) the median	3) 30

❑ THE BOX-AND-WHISKER PLOT

The box plot is a graphical display that describes not only the behavior of the measurements in the middle of the distribution but also their behavior at the ends or tails of the distribution. A box plot graphs a five-number summary of the data: lowest value, lower quartile, median, upper quartile, highest value. Two sets of limits normally placed on the box plot are the inner and outer fences. Recall that the interquartile range, iqr, is the difference in the upper and lower quartiles. Inner fences are located a distance of 1.5 iqr below the lower quartile and 1.5 iqr above the upper quartile. Outer fences are located a distance of 3 iqr above the lower quartile and 3 iqr above the upper quartile. The ends of the box are located at the 25th and 75th percentiles with the median shown as a vertical line through the box.

Horizontal lines called whiskers are drawn from the ends of the box to the smallest and largest data values inside the inner fences.

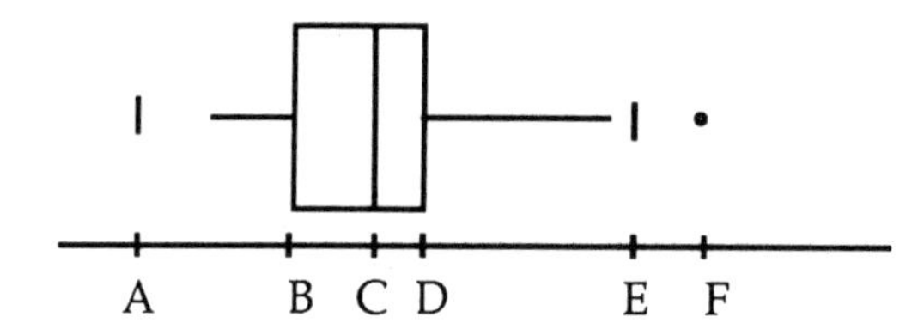

A = lower inner fence
B = 25th percentile
C = median
D = 75th percentile
E = upper inner fence
F = outlier

If the line at the median is at or near the center of the box, this is an indication of symmetry of the data. If one whisker is clearly longer than the other, the data is very likely skewed in the direction of the longer whisker. If you are in doubt as to the length of the whiskers or if you are dealing with small data sets, it is a good idea to compare the mean and median for the data to determine the skewness.

Notice that the box plot gives a graph of summary data since it uses only five statistics and does not show the individual data values as does a stem and leaf plot. We cannot identify the shape of a distribution as well as we can with stem-and-leaf plots or histograms, but we can easily look at relative positions of different sets of data and compare them quite easily. Box plots are extremely useful for very large sets of data.

Program Box plot uses the percentiles calculated by program Percentile and draws a statistical box plot of the data. The calculator range is set at the outer fences as long as no data values are outside those fences. If there are extreme data values (outliers) in either direction, the x-range uses the smallest (or highest) data value for the respective Xmin or Xmax value in the range settings. The program also displays inner fences on the graph. Program Box plot pauses throughout for easier viewing and is resumed by pressing COMP.

PROGRAMMING

Select the AER-II mode. You will see on the display Ø3: TITLE? Type in, using the letters on the right keyboard, SHIFT B o x ⊔ p l o t and press ENT to store the program name. Input the program at the M: prompt as follows:

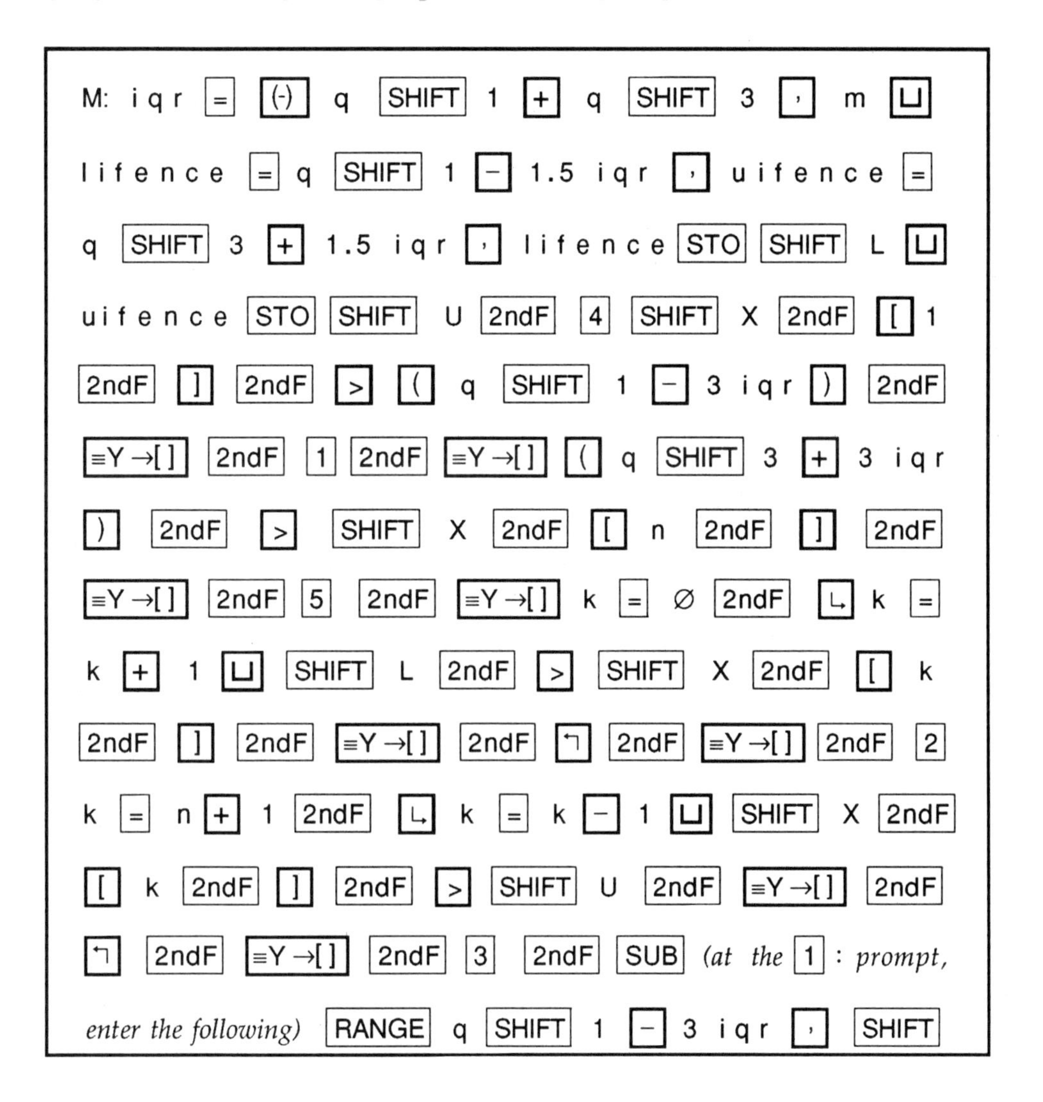

M: i q r = (-) q SHIFT 1 + q SHIFT 3 , m ⊔
l i f e n c e = q SHIFT 1 − 1.5 i q r , u i f e n c e =
q SHIFT 3 + 1.5 i q r , l i f e n c e STO SHIFT L ⊔
u i f e n c e STO SHIFT U 2ndF 4 SHIFT X 2ndF [1
2ndF] 2ndF > (q SHIFT 1 − 3 i q r) 2ndF
≡Y→[ ] 2ndF 1 2ndF ≡Y→[ ] (q SHIFT 3 + 3 i q r
) 2ndF > SHIFT X 2ndF [n 2ndF] 2ndF
≡Y→[ ] 2ndF 5 2ndF ≡Y→[ ] k = Ø 2ndF ↳ k =
k + 1 ⊔ SHIFT L 2ndF > SHIFT X 2ndF [k
2ndF] 2ndF ≡Y→[ ] 2ndF ↰ 2ndF ≡Y→[ ] 2ndF 2
k = n + 1 2ndF ↳ k = k − 1 ⊔ SHIFT X 2ndF
[k 2ndF] 2ndF > SHIFT U 2ndF ≡Y→[ ] 2ndF
↰ 2ndF ≡Y→[ ] 2ndF 3 2ndF SUB *(at the* 1 *: prompt,*
enter the following) RANGE q SHIFT 1 − 3 i q r , SHIFT

Program Box plot is continued on the next page.

X [2ndF] [[] n [2ndF] []] [+] 1 [,] i q r [,] .25 [,] 1.25
[,] 1.25 [2ndF] [SUB] *(at the* [2]*: prompt, enter the following)*
[2ndF] [LINE] [SHIFT] X [2ndF] [[] k [2ndF] []] [,] .75 [,]
q [SHIFT] 1 [,] .75 [DRAW] [⊔] [2ndF] [LINE] q [SHIFT] 1
[,] .5 [,] q [SHIFT] 1 [,] 1 [DRAW] [⊔] [2ndF] [LINE] q
[SHIFT] 1 [,] 1 [,] q [SHIFT] 3 [,] 1 [DRAW] [⊔] [2ndF]
[LINE] q [SHIFT] 1 [,] .5 [,] q [SHIFT] 3 [,] .5 [DRAW]
[⊔] [2ndF] [LINE] q [SHIFT] 3 [,] .5 [,] q [SHIFT] 3 [,]
1 [DRAW] [2ndF] [SUB] *(at the* [3] *: prompt, enter the following)*
[2ndF] [LINE] q [SHIFT] 3 [,] .75 [,] [SHIFT] X [2ndF] [[]
k [2ndF] []] [,] .75 [DRAW] [⊔] [2ndF] [LINE] m [,] .5
m [,] 1 [DRAW] [⊔] [2ndF] [LINE] q [SHIFT] 1 [−] 1.5 i q r
[,] .7 [,] q [SHIFT] 1 [−] 1.5 i q r [,] .8 [DRAW] [⊔]
[2ndF] [LINE] q [SHIFT] 3 [+] 1.5 i q r [,] .7 [,] q [SHIFT]
3 [+] 1.5 i q r [,] .8 [DRAW] [,] [2ndF] [SUB] *(at the* [4]*:*
prompt, enter the following) [RANGE] [SHIFT] X [2ndF] [[] 1 [2ndF]
[]] [−] 1 [SHIFT] X [2ndF] [[] n [2ndF] []] [+] 1 [,] i q r

Program Box plot is continued on the next page.

[,] .25 [,] 1.25 [,] 1.25 [2ndF] [SUB] *(at the [4]: prompt, enter the following)* [RANGE] q [SHIFT] 1 [−] 3 iqr [,] q [SHIFT] 3 [+] 3 iqr [,] iqr [,] .25 [,] 1.25 [,] 1.25 [ENT]

Once the program is entered, your calculator screen should look exactly like the following. Press [2ndF] [▽] to see the contents of the subroutines.

```
M: iqr = -q1 + q3, m ⊔ l
ifence = q1 −1.5 iqr
, uifence = q3 +1.5 i
qr, lifence⇒L⊔uif
ence⇒U [4] X[1]>(q1−
3 iqr) ≡Y→[ [1] ] (q3 + 3
iqr) > X[n] ≡Y→[ [5] ] k
= ∅↳ k= k+1 ⊔ L> X[k]≡
Y→[ ↰ ] [2] k=n+1 ↳ k=k−
1 ⊔ X[k]> U ≡Y→[ ↰ ] [3]
```

```
[1]: RANGE q1 −3 iqr,
X[n]+1, iqr, .25 ,1
.25 , 1.25
```

```
[2]: LINE X[k], .75,
q1,.75 DRAW⊔LINE
q1, .5, q1,1 DRAW ⊔L
INE q1 , 1 , q3, 1 DRA
W⊔LINE q1, .5, q3,
.5 DRAW⊔LINE q3, .
5 , q3, 1 DRAW
```

```
[3]: LINE q3, .75, X[
k], .75 DRAW ⊔ LINE
m, .5, m, 1 DRAW⊔LIN
E q1 −1.5 iqr , .7, q
1−1.5 iqr, .8 DRAW ⊔
LINE q3 + 1.5 iqr, .
7, q3 +1.5 iqr, .8 DR
AW ,
```

4: RANGE X[1]−1, X [n]+1, iqr, .25, 1.25, 1.25	5: RANGE q_1 −3 iqr, q_3 + 3 iqr, iqr, .25, 1.25, 1.25

- Box plots may be drawn vertically as well as horizontally. The program below is designed to draw horizontal box plots because of the dimensions of the calculator screen.

- The trace feature of the Sharp EL-5200 does not work on the box plots constructed by the program given below. Note that the Xscl tic marks on the box plot are set by the program at intervals equal to the interquartile range.

- If you wish to review the values for the lower and inner fences that were computed by the program, press L [=] and U [=]. The other quantities have been stored in lowercase letter variables which are not able to be recalled after the program has finished.

- The box plot will remain on the graphics screen until the range is reset or another graph has been drawn, even if your calculator has been turned off. You may recall the box plot by pressing [2ndF] [T-G-D].

❒ OUTLIERS

Outliers may represent faulty measurements in recording or observation or may be valid measurements which, for one reason or another, differ markedly from the others in the set. Data values that fall between the inner and outer fences are called *mild outliers*. Data values falling outside the outer fences are called *extreme outliers*. Outliers should always be examined. Program Outlier tests for mild and extreme outliers and indicates the outlier(s) by showing the point(s) on the graph. Thus, it should be used *after* executing program Box plot.

PROGRAMMING

Select the AER-II mode. You will see on the display Ø4: TITLE? Type in, using the letters on the right keyboard, SHIFT O u t l i e r and press ENT to store the program name. Input the program at the M: prompt as follows:

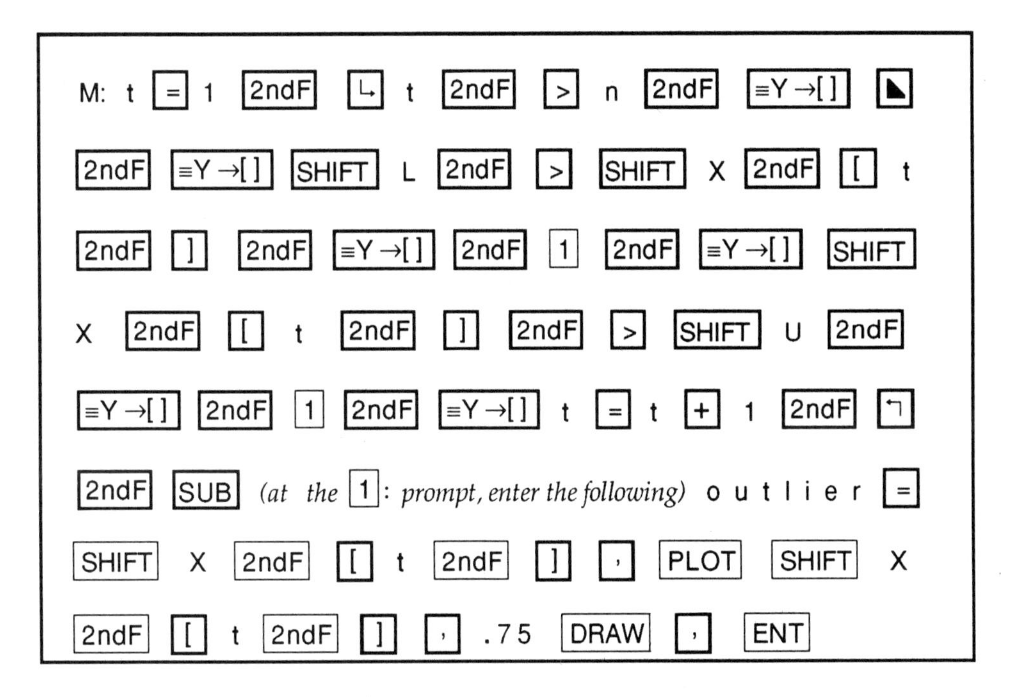

Once the program is entered, your calculator screen should look exactly like the following:

M: t=1 ↳ t> n ≡Y→[◣] L
> X[t] ≡Y→[1] X[t]>
U ≡Y→[1] t=t+1 ↰

Press 2ndF ▽ to see the contents of subroutine 1:

```
[1]: outlier=X[t],P
LOT X[t],.75 DRAW
,
```

- Program Outlier will indicate all values that are widely separated from the rest of the data as long as [COMP] is pressed after each point has been plotted on the graph of the box plot. When you get the message "ANS 6 = " or "ANS 7 = ", you will know all outliers have been found. If you get the message "ANS 1 = n", no outliers have been found.

- Some textbooks use *hinges* rather than quartiles to create the box. Consult your statistics text or your instructor.

EXERCISES

1. The number of times each letter of the English alphabet occurred was counted on one randomly chosen page of a statistics text. The table below lists each of the letters and the percentage that each letter occurred.

A	7.9	H	5.6	O	6.5	V	0.8
B	1.8	I	6.6	P	2.1	W	1.6
C	2.6	J	0.1	Q	0.09	X	0.15
D	3.3	K	0.4	R	6.5	Y	2.4
E	13.7	L	3.7	S	7.3	Z	0.06
F	2.7	M	2.6	T	10.0		
G	2.0	N	8.0	U	1.5		

a) Use program Sort to enter this data and sort it in increasing order.
b) While you are waiting for the calculator to finish, answer the following questions:
   i) What is the most-used letter? the least-used letter?
   ii) What percentage of the letters used are vowels?
   iii) Why do you think the letters T, R, S, L, N and E are the ones most often chosen on the *Wheel of Fortune* game show?
c) Find the quartiles and the median percentage.
d) Construct a box plot of the percentages.
e) Are most of the letters used rarely or frequently? What conclusion can you

draw from the length of the whiskers on your box plot?

f) Identify any outliers.

2. Choose one page at random from your statistics text and one page at random from your English text. (If you choose a page from either with more than one illustration, choose another at random.) Count the occurrence of the letters on each page and calculate the percentage of occurrence of each of those letters. Answer the same questions that are in the problem above for each set of data. Compare the box plots.

3. The following are the numbers of private aircraft which landed at a large metropolitan airport on fifteen consecutive days:

   85 74 67 87 71 89 82 125 73 84 77 82 70 90 38

   Construct a box plot for this data and interpret the graph. Give the five number summary. Find and discuss the possible reasons for any outliers that may be present. You should also consider the practical interpretations of this data and discuss what these numbers would mean to airport controllers, planning for additional runways, etc. Discuss these interpretations and the reasons for the outliers with other students.

ANSWERS

1. b) i) The most used letter is E and the least used letter is Z.
   ii) The letters A, E, I, O and U are used 36.2% of the time.
   iii) These letters are normally chosen since they have the highest frequency of occurrence in most phrases.

   c) There are n = 26 data values. The 25th percentile, q_1, is 1.5. The 75th percentile, q_3, is 6.5. The median, the 50th percentile, is 2.6.

   d) Note that iqr = 5, lower inner fence = -6 and upper inner fence = 14. Tic marks appear at the bottom of the screen and are set 5 units apart. The box plot is

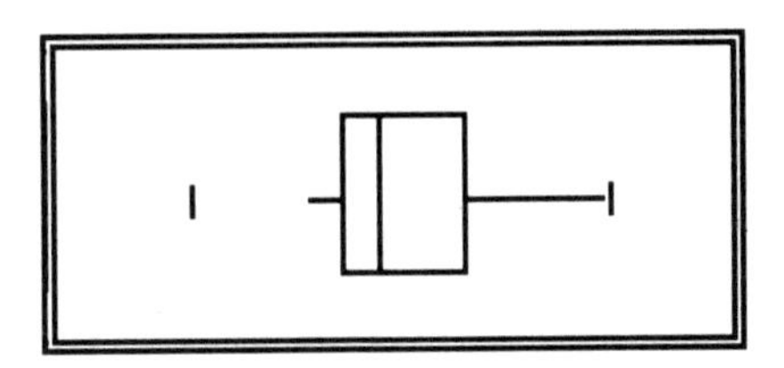

   Notice that the x-axis appears on your calculator screen since the fences involve negative and positive values.

e) Since half of the letters are used about 2.6% of the time or less, most are rarely used. Since the right whisker is much longer than the left, we conclude that the range of the data from q_1 to the lowest data value, 0.06 ,is much less than the range of the data from q_3 to the highest data value, 13.7. Note that the middle half of the data is spread from 1.5 to 6.5. The data seems to be positively (rightward) skewed.

f) There are no outliers in this set of data.

2. In this problem, you should use the calculator to aid you in the construction of the box plots for the two sets of data and then reconstruct them by hand using the same scale on a single set of axes. Look at the lengths of the boxes, the location of the medians and the lengths of the whiskers to determine whether of not the variabilities in the two sets of data are approximately the same or different. Check to see if the pattern of outliers is the same in both data sets if outliers are detected. Note that it does not matter whether or not the number of data values is the same for both groups of data.

3. There are n = 15 data values. The smallest data value is 38, the 25th percentile, q_1, is 71, the median, the 50th percentile, is 82, the 75th percentile, q_3, is 87 and the largest data value is 125.

d) Notice that iqr = 16, lower inner fence = 47 and upper inner fence = 111. Tic marks appear at the bottom of the screen and are set 16 units apart. The box plot is

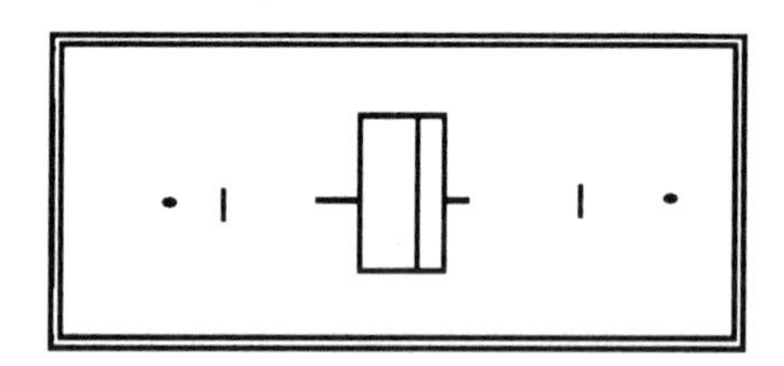

Note that approximately one-half of all the observations lie within the box. Moreover, approximately one-fourth of the observations lie between the median and one side of the box and approximately one-fourth lie between the median and the other side of the box. The slightly longer left whisker is an indication that the data may be negatively skewed.

- Recall that program **Box plot** uses data stored in matrix location X. To use the statistical capabilities of your calculator to find the mean or standard deviation, transfer the data to matrix location S using the procedure described on page 24 of this manual. After transferring the data, use the built in sample mean function to find the mean of this data is 79.60. Since the median of 82 is larger than the mean, this aircraft data is negatively (leftward) skewed.

CHAPTER 3

PROBABILITY

Probability may be regarded as a numerical measure of the chance that a certain outcome of an experiment will occur. A single outcome or any combination of possible outcomes is called an event. Single outcomes which cannot be further decomposed are called simple events. The collection of all possible simple events for an experiment is called the sample space. An understanding of basic probability theory will aid in giving estimates of the reliability of statistical conclusions.

The probability of an event E is a measure of the likelihood of that event and is denoted by P(E). The basic rules of probability are:

- $0 \leq P(E) < 1$ for any event E
- If the sample space is composed of $E_1, E_2, E_3, \ldots, E_n$, then $P(E_1) + P(E_2) + P(E_3) + \ldots + P(E_n) = 1$
- For any events A and B, $P(A \text{ or } B) = P(A) + P(B) - P(A \text{ and } B)$.
- The probability that an event A will not occur, $P(A^c)$, equals $1 - P(A)$.
- The probability that event A will occur given that event B has already happened is $P(A \mid B) = \dfrac{P(A \text{ and } B)}{P(B)}$ provided $P(B) \neq 0$.

Probability also gives the relative frequency with which an event is expected to occur. The larger the probability, the more likely the event is to happen and the smaller the probability, the less likely the event is to occur.

SIMULATION

One approach to determining probabilities experimentally is to perform many repetitions of the experiment (under identical conditions) and determine the proportion of times the event of interest occurs. However, many replications of the experiment may be difficult or impossible to perform. Simulation is the process of representing an experiment with a model. The simulation technique has the advantage over the actual experiment in that many repetitions can be performed quite easily with the aid of a computer or in this case, your calculator.

Two simulations that can easily be done using the calculator are the coin tossing simulation and the die rolling simulation. Once these simulations are performed, you can compare the outcomes of a large number of trials to the "theoretical" results.

Whenever the experiments being simulated involve elements of chance, the simulation techniques are often referred to as Monte Carlo methods. Usually Monte Carlo simulation techniques use random numbers, and most statistics texts include a table of random numbers. These tables are constructed in such a way that every digit has an equal probability of being selected. The Sharp EL-5200 has its own built-in program for generating psuedo-random numbers on the interval [0,1). A "true" random number generator on the interval [0,1) would select each *real* number in that interval with equal probability. Since it is impossible to simulate with the precision of real numbers, random number generators in calculators and computers generate psuedo-random numbers - that is, random numbers that are obtained to a fixed number of decimal places. These random outcomes look and behave like theoretical random numbers.

Three-digit random numbers on the interval [0,1) may be obtained by the random number generation key on the EL-5200. To use this function, place the calculator in the COMP mode (press TAB 3), clear the screen (press CL), and press 2ndF RND. The letters RND should appear on the display screen. Press = several times and you will see some calculator-generator pseudo-random numbers between 0.000 and 0.999.

- Press [TAB] 9. You will notice that the random numbers on your screen all have zeros after the third decimal place. Since this built-in psuedo-random number generator only gives values to three significant digits, there are exactly 1000 possible random numbers that can be generated on the calculator.

❐ COIN TOSS SIMULATION

To simulate the experiment of tossing one fair coin, let the outcome "tails" be represented by Ø and the outcome "heads" be denoted by the number 1. Since RND yields a random number x such that $0.000 \le x < 0.999$, we have $0.000 \le 2x < 1.998$. Notice also that whenever $0.000 \le x < 0.499$, we have $0.000 \le 2x < 0.998$ and that whenever $0.500 \le x < 0.999$, it is true that $1.000 \le 2x < 1.998$.

The keystrokes [2ndF] [INT] will display the integer portion of a number while [2ndF] [FRAC] will display the fractional (decimal) portion of a number. Thus, the function INT(2RND) will display Ø for any value of x between 0 and 0.499 and 1 for any value of x between 0.500 and 0.999. You will probably find it convenient to now press [TAB] Ø since the values displayed will be either the integers 0 or 1.

Key in [2ndF] [INT] 2 [2ndF] [RND], press [=] ten times and observe the results. Record the number of zeros and ones you obtained. What percentage (relative frequency) of zeros and ones did you obtain? Press [=] twenty times more and count the number of zeros and ones generated in the thirty trials. Compute the corresponding percentages. Press [=] twenty times more and count the number of zeros and ones (tails and heads) you obtained in the fifty trials. You should find the relative frequencies of each of the numbers of zeros and ones approaching the theoretical probabilities of the simple events in the sample space {TAIL , HEAD} for the experiment of flipping a fair coin once.

- Note: The first random number that is generated is dependent on the value stored in memory location X. You may simulate randomly choosing a position in a table of random numbers by storing any real number in memory X by pressing the numerical keys corresponding to the real number you choose and [STO] X. Repeat the above coin toss simulation procedure using several different initial values for X and observe the results.

GRAPHING THE RESULTS

A graph of the outcomes of this experiment may help you visually compare the experimental outcome to the theoretical outcome better than looking at a list of numerical results. To perform the simulation and construct a histogram of the results, place the calculator in the STAT mode. Choose the data store mode, clear memory array S and prepare for data entry with the keystroke sequence [TITLE DATA] [2ndF] [CA] [ENT] [2ndF] [T-G-D]. Key in [2ndF] [INT] 2 [2ndF] [RND] and press [DATA]. Notice that the random numbers generated are not displayed on the screen. Instead, you will see "1." indicating that the first data value has been entered into statistical memory array S. Continue pressing [DATA] until the numerical value indicating the number of times you wish to toss the coin appears as the last value on the screen. If you wish to see the actual values that were generated, press [TITLE PRO] and [=]. Scroll through the values with [▲]. Press [TITLE PRO] [2ndF] [T-G-D] to return to the text screen.

Before constructing a histogram of the results, you will probably want to set the range parameters to give you a "good-looking" graph. Since the data values are either 0 or 1, you want two class intervals such that the zeros will fall in the leftmost interval and the ones will fall in the interval on the right. There are many different range setting that will achieve the desired results. For purposes of illustration, set Xmin = 0, Xmax = 2(0.9)= 1.8, Xscl = 1, Δx = 0.9, Ymin = 0, and Ymax = n(0.5) + 5 where n is the number of data values that you have generated.

- Note: The value of Ymax may have to be reset in order to see the top of the highest rectangle. The larger the value of n, the closer Ymax should be to the theoretical value n(0.5).

To construct the histogram, exit the range and press [2ndF] [G(HI)] [DRAW]. Activate the tracing function with [▷] to display the respective number of tails and heads obtained in your experiment.

❑ DIE ROLLING SIMULATION

The sample space for the roll of one fair die is {1, 2, 3, 4, 5, 6}. Since the die is fair, each of these simple events should occur with probability 1/6. To simulate the experiment for the toss of one fair die, we need to generate six distinct integers occurring with equal probability. Consider the following for $0 \leq x < 0.999$:

RND	2RND	6RND	6RND + 1
0.000 - 0.499	0.000 - 0.998	0.000 - 2.994	1.000 - 3.994
0.500 - 0.999	1.000 - 1.998	3.000 - 5.994	4.000 - 6.994

Therefore, the keying sequence needed to produce random integers from 1 to 6 is INT (6 2ndF RND + 1) . With the calculator in the COMP mode, type this keying sequence, press = and observe the results for 10 simulated tosses of a fair die.

GRAPHING THE RESULTS

Rather than observing the numerical values for increasingly larger number of trials, let's graph the outcomes of this experiment as we did for the coin toss simulation. To perform the simulation and construct a histogram of the results, place the calculator in the STAT mode. Choose data store, clear memory array S and prepare for data entry with the keystroke sequence TITLE DATA 2ndF CA ENT 2ndF T-G-D . Key in 2ndF INT (6 2ndF RND + 1) and press DATA . Recall that the random numbers generated are not displayed on the screen. Instead, you see "1." indicating that the first data value has been entered into statistical memory array S. Continue pressing DATA until the numerical value indicating the number of times you wish to toss the coin appears as the last value on the screen.

If you wish to see the actual values that were generated, press TITLE DATA and = . Scroll through the values with △. Press TITLE DATA 2ndF T-G-D to return to the text screen.

Before constructing a histogram of the results, you will probably want to set the range parameters to give you a "good-looking" graph. Since the data values are either 0 or 1, you want two class intervals such that the "zeros" will fall in the leftmost interval and the "ones" will fall in the interval on the right. There are many different range setting that will achieve the desired results. For purposes of illustration, set Xmin = 1, Xmax = 1 + 6(0.9)= 6.4, Xscl = 1, Δx = 0.9, Ymin = 0, and Ymax = n(1 ÷ 6) + 5 where n is the number of data values that you have generated.

- Note: The value of Ymax may have to be reset in order to see the top of the highest rectangle. The larger the value of n, the closer Ymax should be to the theoretical value n(1 ÷ 6).

To construct the histogram, exit the range and press [2ndF] [G(HI)] [DRAW] .

Activate the tracing function with [▷] to display the number of dots that you obtained in your experiment. Remember that the displayed y value is the frequency of the occurrence of either 1, 2, 3, 4, 5 or 6 dots. If you wish the relative frequency, you must divide by the total number of times the die is rolled. The relative frequency should then be compared to the theoretical probability.

EXERCISES

1. Consider the function $y = \sqrt{2x}$ where x is a random number on the interval [0,1).
   a) What are the possible values that INT(y) can assume?
   b) Take the integer portion of y and generate 10 values with the keystroke sequence [2ndF] [INT] [(] [√] [(] 2 [2ndF] [RND] [)] [)] in the STAT mode and enter these as data values with [DATA] . Construct the histogram.
   c) Repeat the experiment for n = 30 and n = 50.
   d) How do the results compare with the simulation that was done earlier using INT(y) for the function y = 2x where x is a random number on the interval [0,1)?
   e) Do you feel that $y = \sqrt{2x}$ could be used instead of y = 2x in the simulation of the toss of one coin? Explain.

2. Answer parts a) through e) in the above problem for INT(y) where y is the function $e^x - 1$. The keystroke sequence for part b) is [2ndF] [INT] [(] [2ndF] [e^x] [2ndF] [RND] [−] 1 [)] .

PERMUTATIONS

Theoretical probabilities are determined on the basis of an ideal experiment that has been repeated infinitely many times. In some experiments it is tedious and inefficient to list all the outcomes in the sample space. The number of simple events is influenced by the method of sampling you use. If you sample *with replacement,* you return the element chosen to the population before you select the next element. If you sample *without replacement,* you do not return the element chosen to the population before choosing the next one. When the *order* in which the elements are listed is important, the order in which the elements are selected must be considered. (For instance, on tossing a coin two times, HT is a different simple event than TH since a difference face appears as a result of each toss). If the order in which the elements are chosen is not important, the order in which those elements are selected should not be considered. (For instance, if three people are to be chosen from a group of 8 people to form a committee, the committee is the same regardless of how the people chosen are arranged.)

A *permutation* is an ordered arrangement without repetition of elements of a set of distinct objects. The notation for the number of permutations of n different objects taken r at a time is $nPr = n!/r!$. For the set {a,b,c}, the permutations are ab, ba, ac, ca, bc, and cb with $3P2 = 6$.

The Sharp EL-5200 has a built-in function to evaluate nPr. To find 3P2, place the calculator in the COMP mode, press [TAB] 0 (since nPr is always a whole number) and press 3 [2ndF] [nPr] 2 [=] to obtain the value of 6.

COMBINATIONS

A *combination* is a selection of the distinct objects of a set without regard to order. As with permutations, repetitions of elements are not allowed. The essential difference between permutations and combinations is that combinations ignore the order in which the objects are chosen. The notation for the number of combinations of n different objects chosen r at a time is $nCr = \binom{n}{r} = \frac{n!}{r!\,(n-r)!}$ and nCr is also called the binomial coefficient. The combinations are ab, ac, and bc (or ba, ca, and cb) for the set {a,b,c} with 3C2 = 3. The calculator has a built-in function to evaluate nCr. To evaluate 3C2, place the calculator in the COMP mode, press [TAB] 0 (since nCr is also a whole number) and press 3 [2ndF] [nCr] 2 [=] to obtain 3.

EXERCISES	ANSWERS
1) Find 18 P 7.	1) 160392960
2) Compute 20 C 8.	2) 125970
3) How many different 4-digit telephone numbers can be formed from the digits 1 through 9 if each digit can be used only once in each number?	3) 3024
4) A student must choose 3 electives from 7 different courses being offered. In how many ways can she make her choice?	4) 35

5) Find $\binom{69}{61}$.

5) 8,361,453,672

6) Try to compute 70P35 and 100C50. What happens?

- To clear the Sharp EL-5200 from an error message, press the red CL key.

Please recall that you are not to bend back the soft vinyl cover of the calculator. However, the left keyboard of the calculator itself will flip over to the right. The type of error you have made is explained by number on the back of the left keyboard of the calculator. (A program that can be used to evaluate combinations for values of $n \geq 70$ will be given in Chapter 4.)

CHAPTER 4

DISCRETE RANDOM VARIABLES

MEANS AND VARIANCES OF DISCRETE PROBABILITY DISTRIBUTIONS

The mean and variance for a discrete random variable x with probabilities p(x) are given, respectively, by the formulas

$$\mu = \sum_{\text{all } x} x\, p(x) \quad \text{and} \quad \sigma^2 = \sum_{\text{all } x} x^2\, p(x) - \mu^2$$

Recall that the Sharp EL-5200 will compute means and variances for data input in the STAT mode. As the data is input, the calculator counts the data values that are input beginning with 1 and ending with n where n is the total number of data values. The mean of the x data is determined by the formula $\bar{x} = \frac{\sum x}{n}$ and the built-in mean function obtained by pressing [2ndF] 4 can be used to find the mean of a probability distribution. A similar argument holds for using the built-in standard deviation function of the calculator. Recall that whenever the data represents a sample chosen from a population, the key for s_x ([2ndF] 5) should be used to calculate the standard deviation. Whenever the data represents the population itself, as in the case of a probability distribution, the key for obtaining the standard deviation is σ_x ([2ndF] 6).

If you do not wish to save your data and if you finish working with your data before you change modes or the calculator is turned off, you can use the non-store mode to readily compute means and variances for any type of data. For example, suppose you wish to use the built-in functions in the STAT mode to compute the mean and variance of the following probability distribution for the discrete random variable x:

x	0	1	3	5
p(x)	.20	.15	.35	.30

To find the mean and variance of this probability distribution using the built-in functions in the calculator, follow the steps given below:

- Place the calculator in the STAT mode and choose 2 for the non-store mode.
- Enter the data values as x [X] f where f is the numerator of each p(x) when the p(x) value is written as a fraction with denominator of 100. Press [DATA] (the [M+] key) after each entry. The keystrokes to enter the values for the probability distribution given above are 0 [X] .20 [DATA] 1 [X] .15 [DATA] 3 [X] .35 [DATA] 5 [X] .30 [DATA]. Note that the number on the right of your screen should read 1.00 when you have finished entering your data values since the sum of the probabilities for any probability distribution is 1.
- To obtain the mean μ of this random variable press [2ndF] 4 and obtain 2.700.
- To obtain the standard deviation σ, press [2ndF] 6 and obtain 1.8735. If you wish the variance σ^2 press [2ndF] 6 [x^2] [=] to obtain 3.5100.

A problem arises, however, if you wish to save the data or draw a histogram of the data when you are using the non-store mode because you are not entering the data into the statistics S array. The built-in histogram program in the STAT mode will graph only frequency distributions and the values of p(x) are not whole numbers. This problem can be remedied, however, by rewriting the formula for μ as $\sum x \frac{f}{100} = \frac{\sum xf}{100}$ where f can be considered the number of times out of 100 that the value of x occurs. Thus, by entering the values of x as xf where f is the numerator of p(x) when each p(x) is written as a fraction with denominator of 100, you can draw a histogram of the probability distribution using the [G(HI)] key and the appropriate range settings. Data may be stored into another matrix location if you wish it saved for future reference.

To use the data store mode to find the mean and variance of the probability distribution given on the previous page , follow the steps given below:

- Place the calculator in the STAT mode and choose 1 for data store.
- Clear the S data array by pressing [DATA] (TITLE) [2ndF] [CL] and [ENT].
- Return to the text screen using [2ndF] [T-G-D] .
- Enter the data values as x [×] f where f is the numerator of each p(x) when the p(x) value is written as a fraction with denominator of 100. Press [DATA] (the [M+] key) after each entry. The keystrokes to enter the values for the probability distribution given above are 0 [×] 20 [DATA] 1 [×] 15 [DATA] 3 [×] 35 [DATA] 5 [×] 30 [DATA]. Note that the number on the right of your screen should read 100 when you have finished entering your data values. Press [2ndF] 4 to obtain the mean μ and [2ndF] 6 to obtain σ.

BINOMIAL PROBABILITIES

The binomial probability distribution is one of the most widely used discrete probability distributions. The four characteristics of a binomial experiment are:

- The experiment consists of a sequence of n identical trials (repetitions).
- There are only two possible outcomes of each trial, referred to as "success" and "failure".
- The probability of success on each trial, denoted by p, remains constant from trial to trial.
- The trials are independent.

Letting the random variable x be the number of successes occurring in n trials, the binomial probability distribution gives the probabilities associated with this discrete random variable. The binomial probability function can be used to compute the probability of x successes in n trials for any binomial experiment. This function is $p(x) = \binom{n}{x} p^x (1 - p)^{n - x}$ for $x = 0, 1, 2, \ldots, n$ and $\binom{n}{x} = nCx$. The following program makes use of the built-in combinations function in the EL-5200 and will calculate values of the binomial probability function as long as n, the number of trials, is less than 70.

PROGRAMMING

Select the AER-I mode by sliding the switch on the left of the calculator to the top position. You will see on the display Ø5: TITLE?. Type in, using the letters on the right keyboard, B I N O M I A L and press ENT to store the program name. Input the program at the M: prompt as follows:

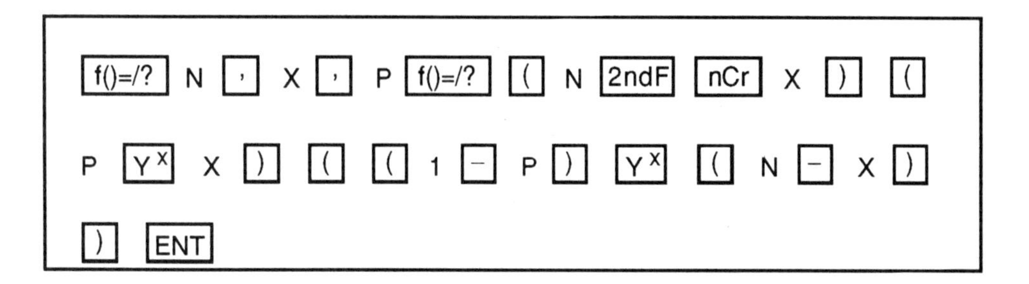

Once the program is entered, your calculator screen should look exactly like the following:

M: f(N,X,P) = (N C X)
(P Yx X) ((1−P) Yx (N
−X))

- If your calculator screen has a different number appearing in the space for the next program entry than that in the program in this manual, you have entered the programs in a different order. This will in no way affect the execution of the program.

PROGRAM EXECUTION

To execute or "run" any program, place the calculator in the COMP mode. Press TITLE PRO until the name of the desired program appears. Press COMP and key in the values of n, x, and p as the calculator asks for them, pressing COMP after each one. After the last value has been keyed in and COMP pressed, the value of the binomial probability will appear on the right-hand side of the screen.

(For the following exercises, press TAB 4 if you wish to get the value under "answer".)

EXERCISES	ANSWERS
1) Find the binomial probability p(x) for n = 10, x = 4, and p = .3.	1) .2001
2) A store manager has noticed that 43% of all purchases are made with credit cards. He randomly selects 57 purchases during a given day and notes the method of payment. What is the probability that exactly 25 of these purchases were made using a credit card?	2) .1051
3) A fair coin is tossed four times. Let x be the number of tails observed on the four tosses. Give the probability distribution for x in tabular form.	3) see table below

x	p(x)
0	.0625
1	.2500
2	.3750
3	.2500
4	.0625

- Once you initially use the PRO key to call the program BINOMIAL, press the COMP key and enter n = 4, x = ∅, and p = .5, these values are stored in memories N, X, and P. Thus, it is not necessary to reenter the values of n and p as they do not change when finding the above probability distribution. Once the initial values are entered and p(0) = .0625 is found, simply press COMP and when the prompts "N=? " and " P=? " appear, press COMP again to reuse the values of n = 4 and p = .5. If you wish to check which value is stored in memory when a prompt appears, (i.e., N=?) , press PB and the value that is stored will appear on the display screen.

4) Find the probability of obtaining 34, 35, 36, or 37 successes in a binomial experiment consisting of 41 trials for which the probability of success is 5/7.	4) .0668

❑ INDEPENDENT ACCESSIBLE MEMORY

The Sharp EL-5200 has four types of memories for storing numerical data. The independent accessible memory, the memories for constants (store memories for variables A - Z), the memory for array variables used in the MATRIX and STAT modes, and the memory for flexible variables used in the AER-II mode.

The independent accessible memory (IAM) is quite useful in calculations such as exercise 4 above. Before starting a calculation using the IAM, you must clear the contents of store memory M by pressing [CL] and [⇒M] keys. To store the result of a calculation in memory, press [⇒M] after performing the computation. To add the result of another calculation to the memory contents, press [M+]; to subtract the result of a calculation from the memory contents, press [2ndF] [M+]. To recall the present memory contents, press [RM].

If you wish to use the result of a previous calculation in a program, you may store the result in the IAM (using [⇒M]). When you are in either the AER-I or AER-II programming modes, press [RM] at the point in the program where you wish the result to appear. If you use this feature, be careful to not use the variable M in your program to represent a different value as whatever you store in the IAM is stored in memory location M.

To use the IAM in exercise 4 in the previous set of exercises, first clear the contents of memory M. Find p(34) using the program BINOMIAL and store the result in memory using [⇒M]. Press [COMP] and find p(35). Add the result of this calculation to memory using [M+] . After finding p(36) and p(37) and storing these results in memory, press [RM] to find the probability that x equals 34, 35, 36, or 37.

Also note that it is not necessary to use a decimal approximation for the probability of success (5/7) since 5 [÷] 7 can be entered directly into the calculator for greater accuracy at the "P=?" prompt.

■ CALCULATION OF BINOMIAL PROBABILITIES WHEN THE NUMBER OF TRIALS IS MORE THAN 70

Recall that the built-in calculator function to evaluate combinations fail when $n \geq 70$. This problem can be solved by mathematically rewriting the formula for nCx to avoid division of large factorials. The resulting program also uses the fact that nCx = nCn-x whenever $x > n/2$ so that less iterations of this formula are required resulting in increased computational speed. This program has been named BINOMIAL 7∅ to distinguish it from the preceding formula that was given for the calculation of binomial probabilities. You will notice that BINOMIAL 7∅, which can be used to find binomial probabilities for any value of n, requires slightly more computational time than BINOMIAL which works only for $n < 70$. If time is a factor, use the appropriate program.

- If you wish to look at the flow chart and mathematical derivation of the formula used in the program B I N O M I A L 7∅, consult Appendix I at the back of this manual.

- The following program uses the logical true and false operators. When you first press the key [≡Y →[]], the symbols "≡Y→[" appear on the calculator screen. The second time this key is pressed, if you have not exited the program in the meantime, the symbol "] " appears on the screen. If when checking your program, you see " ≡Y→[" instead of "] " on your screen, you will have to edit this line by pressing [≡Y →[]] and keystroking over this again to produce the symbol "] ".

PROGRAMMING

Select the AER-I mode. You should see on the display Ø6: TITLE?. Type in B I N O M I A L [⊔] 7 Ø and press [ENT] to store the program name. Input the main routine at the M: prompt as follows:

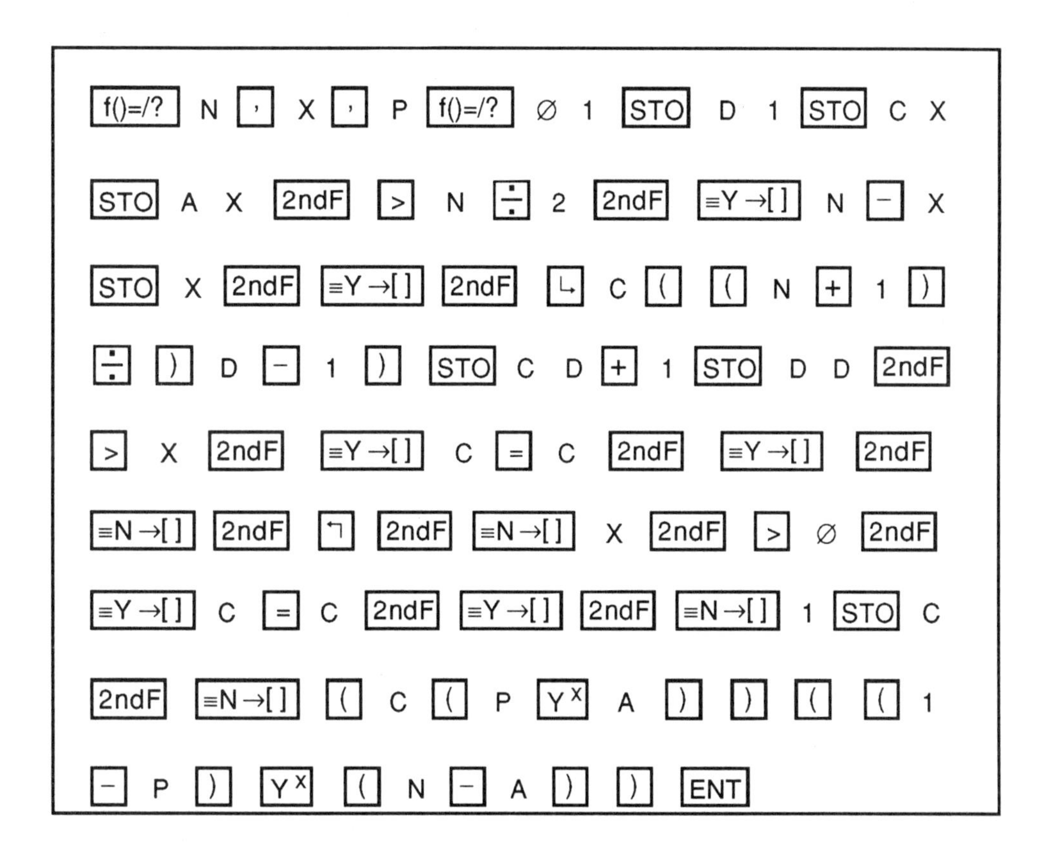

Once the program is entered, your calculator screen should look exactly like the following:

```
M: f(N,X,P)=Ø1⇒D1
⇒CX⇒AX> N÷2 ≡Y→[N−
X⇒X] ↳ C((N+1)÷ D−1
)⇒CD+1⇒D D>X≡Y→[C
=C] ≡N→[ ↰ ]X> Ø ≡Y→[
C=C] ≡N→[1⇒C] (C (P
Y^X A) ((1−P) Y^X (N−
A))
```

• Note: Whenever the calculator is performing a "lengthy" computation, you will see ← and → flashing in the bottom corners of the display screen.

EXERCISES	ANSWERS
1) Find the binomial probability p(x) for $n = 70$, $x = 26$, and $p = .38$.	1) .0972
2) A store manager has noticed that 43% of all purchases are made with credit cards. He randomly selects 57 purchases during a given day and notes the method of payment. What is the probability that exactly 25 of these purchases were made using a credit card?	2) .1051
3) Find the binomial probability p(x) for $n = 100$, $x = 91$, and $p = .93$.	3) .1040
4) Find the probability of obtaining either 75 or 76 successes in a binomial experiment consisting of 100 trials for which the probability of success on any trial is 2/3.	4) .0295

• Most textbooks include tables of binomial probabilities that can be easily used to find certain values of the binomial probability function p(x). However, these tables are not complete in that they do not include all values of n and p. Doing the calculations by hand for the values of n and p not included in these tables can be very tedious, even with a scientific calculator. Because of this table limitation, most statistics texts also include a section on approximating the binomial distribution with the normal probability distribution for large values of n. The above program eliminates the need for binomial tables and approximations.

SIMULATING THE BINOMIAL EXPERIMENT OF TOSSING FOUR FAIR COINS

Suppose that four identical fair coins are tossed and the number of heads (or tails) is counted. The function INT(2RND) + INT(2RND) + INT(2RND) + INT(2RND) will produce the values 0, 1, 2, 3, and 4. Why?

The following program will simulate the toss of four fair coins n times with the number of heads being recorded for each of the n tosses. Data is stored in memory array S so that a histogram can be drawn (in the STAT mode) of the results of the n tosses.

This program makes use of the EL-5200's conditional expression judgment function. Conditional jumps use the [≡Y →[ ]] (yes) and [≡N →[ ]] (no) keys. When a condition specified in the program is met, the calculator performs the operation enclosed in the [ ] preceded by "≡Y→". If the condition is not met, the program is directed to the operation enclosed in the [ ] preceded by "≡N→". This program also uses the looping function which allows the same calculation to be repeated many times. A loop start point begins with [↱] and [↰] directs the program to repeat the calculations following the ↳ character.

PROGRAMMING

Select the AER-II mode. You should see on the display Ø7: TITLE?. Type in, using the letters on the right keyboard, 4 [⊔] c o i n [⊔] t o s s and press [ENT] to store the program name.

- Up to 61 characters may be used in a single program title. The number preceding TITLE? will depend on the order in which you enter the programs in your calculator.

Input the main routine at the M: prompt as follows:

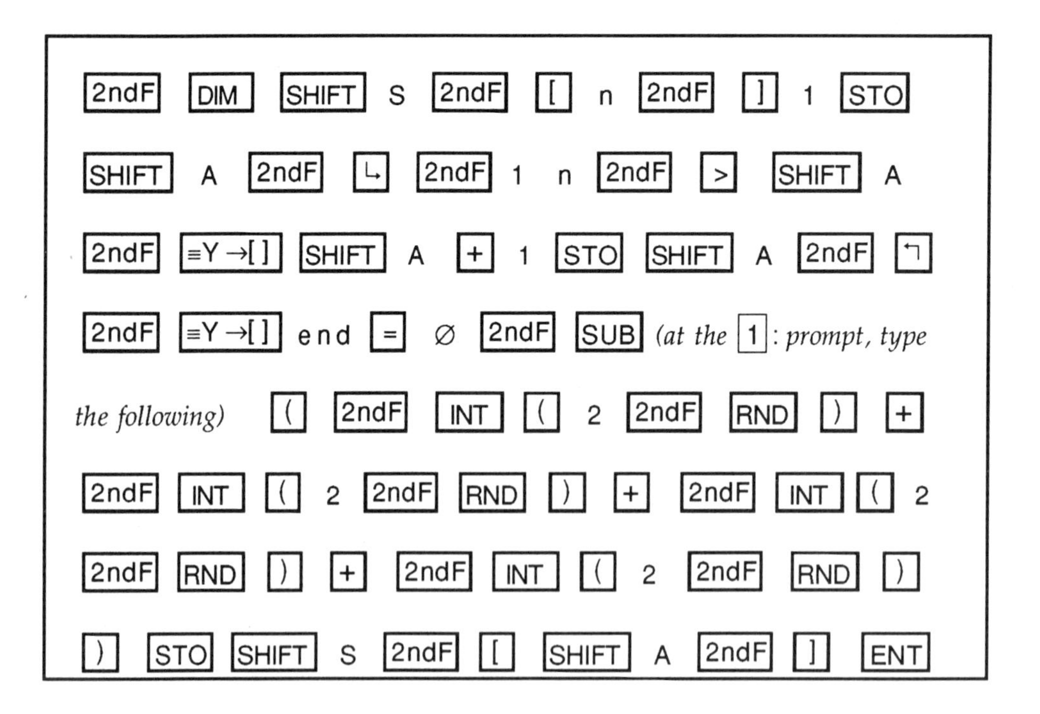

[2ndF] [DIM] [SHIFT] S [2ndF] [[] n [2ndF] []] 1 [STO]
[SHIFT] A [2ndF] [↳] [2ndF] 1 n [2ndF] [>] [SHIFT] A
[2ndF] [≡Y→[ ]] [SHIFT] A [+] 1 [STO] [SHIFT] A [2ndF] [↰]
[2ndF] [≡Y→[ ]] e n d [=] Ø [2ndF] [SUB] *(at the* [1]: *prompt, type*
the following) [(] [2ndF] [INT] [(] 2 [2ndF] [RND] [)] [+]
[2ndF] [INT] [(] 2 [2ndF] [RND] [)] [+] [2ndF] [INT] [(] 2
[2ndF] [RND] [)] [+] [2ndF] [INT] [(] 2 [2ndF] [RND] [)]
[)] [STO] [SHIFT] S [2ndF] [[] [SHIFT] A [2ndF] []] [ENT]

Once the program is entered, your calculator screen should look exactly like the following:

```
M: DIM S[n] 1⇒A ↳ [1] n
>A ≡Y→[ A +1⇒ A ↰ ] e n d
=∅
```

- When you first press the key [≡Y →[ ]], the symbols "≡Y→[" appear on the calculator screen. The second time this key is pressed, if you have not exited the program in the meantime, the symbol "] " appears on the screen. If when checking the second line of your program with the above, you see

  ">A ≡Y→[A +1⇒ A ↰ ≡Y→[e n d "

  on your screen, you will have to edit this line by pressing [≡Y →[ ]] and keystroking over this again to produce the symbol "] ".

Press [2ndF] [▽] to see the contents of subroutine 1 which should appear on your screen exactly like this:

```
[1]: (INT (2RND ) + I
NT (2RND ) + INT (
2RND ) + INT (2RND
 ) )⇒S[A]
```

Recall that statistical graphs are drawn with the calculator in the STAT mode and that in the data store mode all data entered for a statistical calculation are stored in memory S. Array memory S is write-protected in that new data values are not automatically written over values already stored in S. Thus, whenever memory array S is used, it must be cleared manually or an error message ("error 5") will result.

- Statistical memory S must be cleared before drawing a histogram each time the procedure is used.

RUNNING THE PROGRAM

To display in histogram form the values generated by our modification of the function RND for the experiment of tossing four fair coins, we do the following:

- With the calculator in the COMP mode, press [TAB] Ø since the values generated will be the whole numbers 0, 1, 2, 3, and/or 4. (Be certain that FIX is darkened on the display screen.)

- If you are not in the STAT mode, use the slide switch to choose the STAT mode. Press 1 to enter the DATA STORE mode. Clear memory S as instructed above.

- Return the calculator to the COMP mode. Press TITLE [PRO] until you see the program title 4 coin toss . Press [COMP] and enter the value for n, the number of times you wish the four coins to be tossed. Press [COMP] and wait until you see the message "end = 0".

- Place the calculator in the STAT mode. Press 1 (for data store mode). You should see "S: [1,n], P " on the screen with the value you chose for n displayed.

- Press [RANGE] to set the range parameters for drawing your histogram. The range parameters determine the settings for the dimensions of the graphics screen. When you press [RANGE] to display the current settings, the blinking cursor appears over the first value, Xmin. To change its value, type in the new setting and press [=]. Use [▽] to move to the next setting if necessary.

You will probably get the "best-looking" picture by setting Xmin = 0, Xmax = 4.5 , Δx = 0.9 , and Ymin = 0. The value for Ymax will vary with your setting for n, the number of times the four coins are tossed. You can obtain a good estimate of the value for Ymax by using Ymax = n P(2) = n (.3750). Why?

• Press RANGE again to return to the text screen. Press 2ndF G(HI) DRAW and the histogram will appear on the graphics screen. (You may need to reset the value of Ymax to a larger value if the "top" of the largest rectangle is off the screen.)

• If you wish to see the data that was generated by the program 4 coin toss, you may display the values that have been stored in memory array S using the procedure described below.

❐ DISPLAYING VALUES IN THE STORE MEMORIES

If you wish to see which values the calculator has randomly generated, press TITLE/DATA and position the blinking cursor over the S. Press =/SET and the data values will appear on the right-hand side of the display in the reverse order in which they were entered. Press and hold △ to scroll through the values. Press TITLE/DATA Return to the text screen with 2ndF T-G-D .

• Recall that changing a setting in the range always clears the graphics screen. However, if you do not plan to change one of the range settings before drawing a histogram, you must clear the graphics screen. After drawing the histogram, press 2ndF G.CL while in the STAT mode before proceeding to the generation of data for another value of n. If you do not, the histogram for your new value of n will be drawn on top of the one you had previously. Also remember to clear memory array S before generating data for a new value of n or you will get an error message while running the program 4 coin toss.

EXERCISES

1) Let n = 10 and draw the resulting histogram.

2) Let n = 50 and draw the resulting histogram.

3) Let n = 100 and draw the resulting histogram. Compare these results with the histogram for the theoretical probability distribution of the toss of 4 fair coins. Does it appear visually that as n increases, the shape of your histograms are becoming closer to the shape of the theoretical (binomial) probability distribution?

❐ EXPLORATION

We have seen that subroutine [1] in the program 4 coin toss generates random numbers between and including 0 and 4, simulating the number of heads obtained in the toss of four fair coins n times. When these results were graphed in the form of a histogram, the shape of the histogram approached the shape of the histogram for the theoretical probability distribution of the number of heads in tossing the 4 coins.

Recall that the function RND generates random numbers x such that $0 \leq x < .999$. Thus, 5RND will generate random numbers x such that $0 \leq x < 4.995$ and the command INT(5RND) will generate random numbers between and including 0 and 4. Could the expression INT(2RND) + INT(2RND) + INT(2RND) + INT(2RND) in the program 4 coin toss be replaced by INT(5RND) and achieve the same results?

Enter a new program into your calculator, using the AER-II mode. When the prompt Ø8: TITLE? appears on the screen, press [SHIFT] U n i f o r m and press [ENT] to store the program name. The program Uniform is the same as the program 4 coin toss except for the subroutine. Thus, at the M: prompt, enter the same program as you did for 4 coin toss (see page 72), but in place of subroutine [1] in the program 4 coin toss, enter the following as subroutine [1] to obtain the program Uniform:

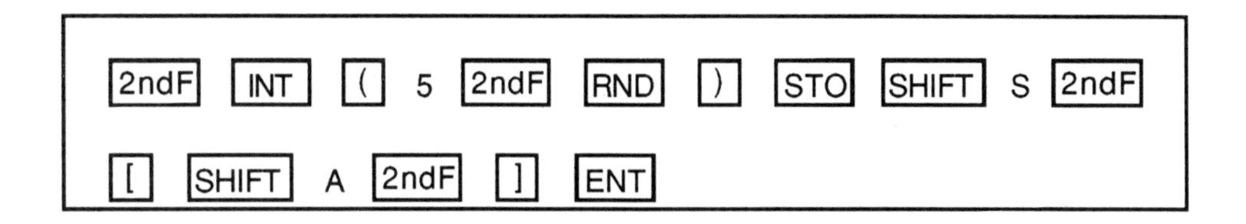

Once the program is entered, your calculator screen should look exactly like the program 4 coin toss in the main program (see page 72) and subroutine 1 should appear as:

```
[1]: INT (5 RND )⇒S[
A]
```

The program Uniform is executed in the same manner as the program 4 coin toss. (Be sure to clear memory array S before using the program or an error will result.) Note that if we could replace the subroutine in 4 coin toss with the subroutine in Uniform, we would expect to obtain histograms of similar shape when performing the experiment. Let's see what happens.

- Before drawing any of the following histograms, set the following range parameters: Xmin = 0, Xmax = 4.5, $\Delta x = 0.9$, Ymin = 0, Ymax ≈ (n)(.375).

1) Using the program 4 coin toss, let n = 50, 100, and 200. In each case, draw what you consider to be the "best-looking histogram" on your calculator screen. Copy those histograms onto the graphs on page 79. (Use the tracing function to obtain the heights of the rectangles.)

2) Using the program Uniform, let n = 50, 100, and 200. In each case, draw what you consider to be the "best-looking histogram" on your calculator screen. Copy those histograms onto page 80. (Use the tracing function to obtain the heights of the rectangles.)

3) Compare your histograms obtained in 1) and 2). Do you believe that you could use the program Uniform in place of the program 4 coin toss to simulate the binomial experiment of the toss of 4 fair coins? Give your answer to this question and the reason(s) for that answer. (Further explanation of the answer to this question will be given in Chapter 6.)

HISTOGRAMS FOR THE PROGRAM 4 COIN TOSS

n = 50

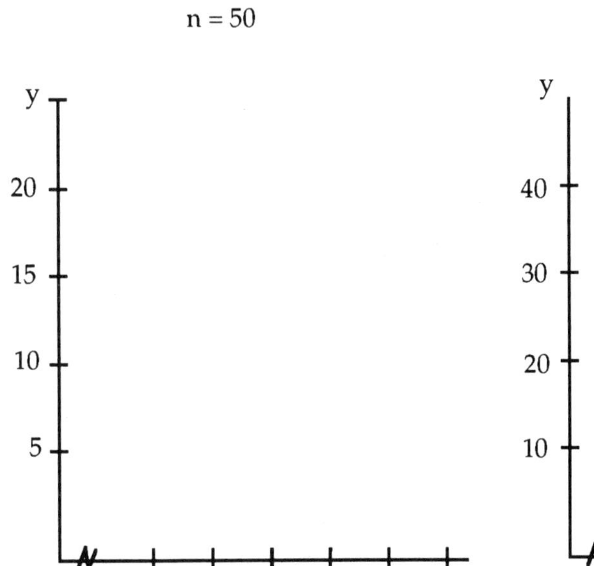

n = 100

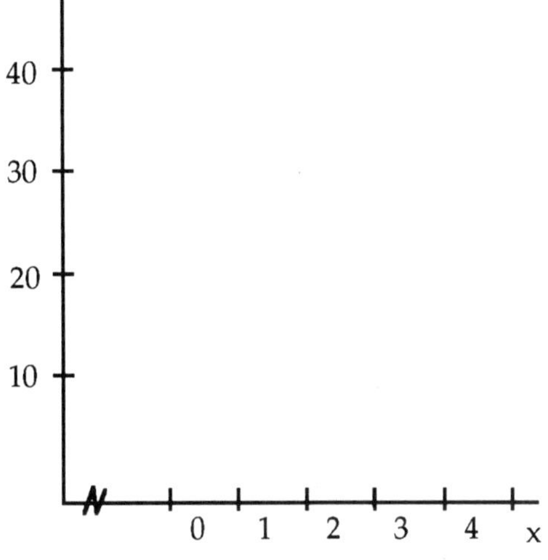

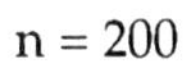

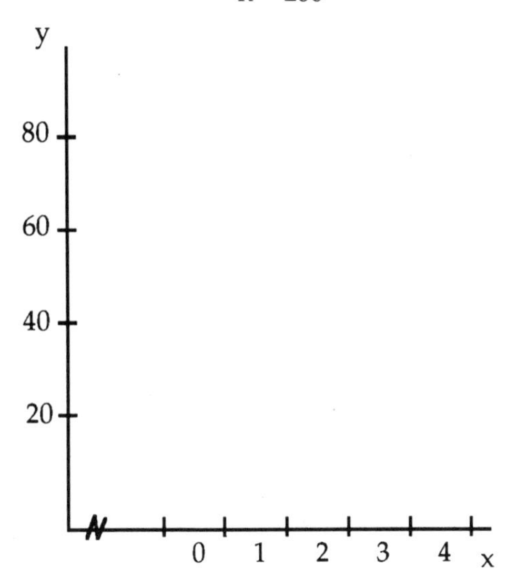

HISTOGRAMS FOR THE PROGRAM UNIFORM

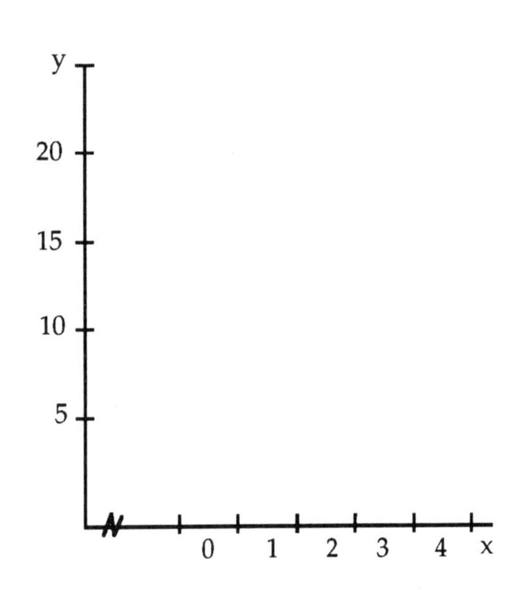

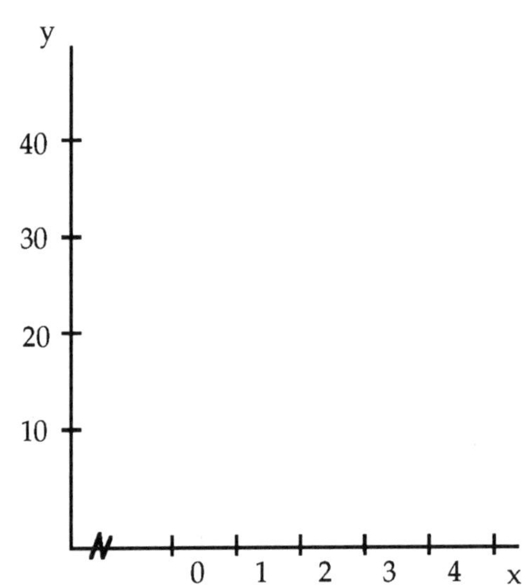

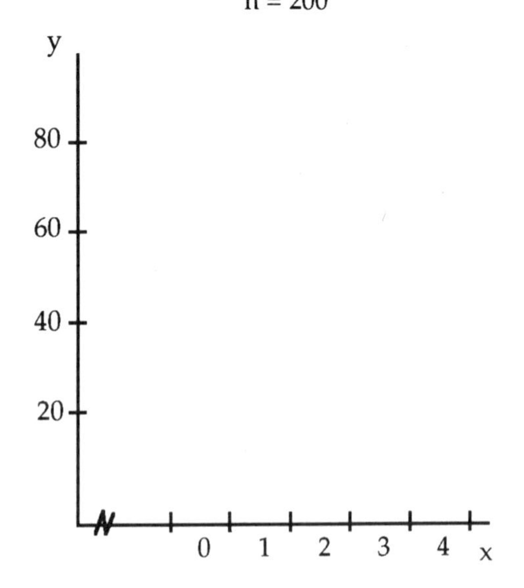

GRAPH OF THE BINOMIAL DISTRIBUTION

The binomial probability distribution varies with different values of n, the number of trials, and p, the probability of success on any one trial. The following program will show the binomial distribution probability histogram for $n < 70$ and any value of p between 0 and 1. The program in the preceding section stores data to the S matrix used by the STAT mode of the calculator. This program stores the binomial probabilities in matrix location B and uses the LINE function to construct the histogram. It does not use any of the built-in statistical functions or graphs.

The following graphs are examples of those produced using the program Binomial dist graph.

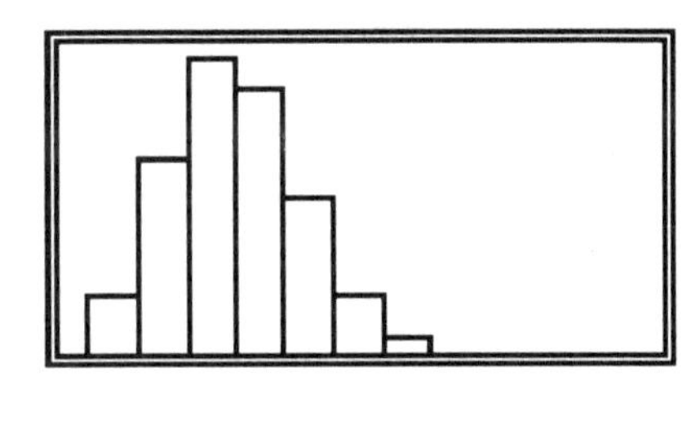

$n = 10, p = 0.25$

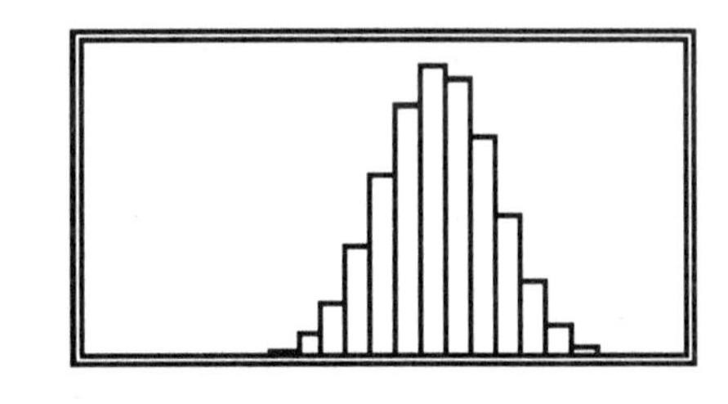

$n = 21, p = 0.58$

PROGRAMMING

Select the AER-II mode and you will see on the display Ø9: TITLE?. Type in, using the letters on the right keyboard, [SHIFT] B i n o m i a l [␣] d i s t [␣] g r a p h and press [ENT] to store the program name. Input the program at the M: prompt as follows:

[2ndF] [DIM] [SHIFT] B [2ndF] [[] n [+] 1 [2ndF] []] n [STO]

[SHIFT] N Ø [STO] [SHIFT] X [2ndF] [↳] [2ndF] [1] [(]

[SHIFT] N [+] 1 [)] [2ndF] [>] [SHIFT] X [2ndF] [≡Y →[ ]]

[2ndF] [↰] [2ndF] [≡Y →[ ]] Ø [STO] [SHIFT] H [SHIFT] H [+] 1

[STO] [SHIFT] H [SHIFT] B [2ndF] [[] [SHIFT] H [2ndF] []]

[2ndF] [>] [SHIFT] B [2ndF] [[] [SHIFT] H [+] 1 [2ndF] []]

[2ndF] [≡N →[ ]] [2ndF] [↰] [2ndF] [≡N →[ ]] [2ndF] [4] [2ndF] [2]

Ø [STO] [SHIFT] C [2ndF] [↳] [2ndF] [3] [SHIFT] C [2ndF] [≠]

[(] [SHIFT] N [+] 1 [)] [2ndF] [≡Y →[ ]] [2ndF] [↰] [2ndF]

[≡Y →[ ]] [2ndF] [SUB] *(at the* [1]: *prompt, enter the following)* [SHIFT]

N [2ndF] [nCr] [SHIFT] X [(] p [Y^X] [SHIFT] X [)] [(]

[(] 1 [−] p [)] [Y^X] [(] [SHIFT] N [−] [SHIFT] X [)]

[)] [STO] [SHIFT] B [2ndF] [[] [SHIFT] X [+] 1 [2ndF] []]

[SHIFT] X [+] 1 [STO] [SHIFT] X [2ndF] [SUB] *(at the* [2] :

prompt, enter the following) [RANGE] [(-)] 1 [,] [SHIFT] N [+] 1

Note: The program **Binomial dist graph** is continued on the next page.

[,] 1 [,] Ø [,] [SHIFT] B [2ndF] [[] [SHIFT] H [2ndF] []]

[,] [SHIFT] B [2ndF] [[] [SHIFT] H [2ndF] []] [÷] 5 [2ndF]

[SUB] *(at the* [3] : *prompt, enter the following)* n [2ndF] [nCr]

[SHIFT] C [(] p [Y^x] [SHIFT] C [)] [(] [(] 1 [−] p [)]

[Y^x] [(] [SHIFT] n [−] [SHIFT] C [)] [)] [STO] [SHIFT] U

[SHIFT] C [−] .5 [STO] [SHIFT] A [SHIFT] C [+] .5 [STO]

[SHIFT] B [2ndF] [LINE] [SHIFT] A [,] [SHIFT] U [,]

[SHIFT] B [,] [SHIFT] U [DRAW] [2ndF] [LINE] [SHIFT] A

[,] [SHIFT] U [,] [SHIFT] A [,] Ø [DRAW] [2ndF] [LINE]

[SHIFT] B [,] [SHIFT] U [,] [SHIFT] B [,] Ø [DRAW]

[SHIFT] C [+] 1 [STO] [SHIFT] C [2ndF] [SUB] *(at the* [4] :

prompt, enter the following) [2ndF] [SYMBOL] 6 [=] n [×] p

[,] s i g m a [=] [√] [(] n [×] p [×] 1 [−] p [)]

[)] [,] [2ndF] [SYMBOL] 6 [STO] [SHIFT] K [␣] s i g m a

[STO] [SHIFT] L [ENT]

Once the program is entered, your calculator screen should look exactly like the following. Recall that you press [2ndF] [▽] to view the contents of the subroutines.

```
M: DIM  B[n+1] n⇒N ∅
⇒X ↳ [1] (N+1) > X ≡Y→[ ↰
] ∅⇒ H ↳ H+1⇒ H B[H] > B
[H+1] ≡N→[ ↰ ] [4] [2] ∅⇒C
↳ [3] C ≠ (N+1) ≡Y→[ ↰ ]
```

```
[1] :  N C X (p Y^x X) ((1-p
) Y^x  (N-X) ) ⇒B[X+1]
X+1 ⇒X
```

```
[2] : RANGE  -1, N +1, 1
, ∅, B[H] ,B[H] ÷ 5
```

```
[3] :  n C C (p Y^x C) ((1-p
) Y^x (n-C) )⇒U C-.5⇒
A C +.5⇒B LINE   A ,U,
B , U DRAW LINE   A , U,
A , ∅ DRAW LINE   B, U,
B , ∅ DRAW  C +1⇒C
```

```
[4] : μ=nxp,sigma=√ (
nxpx(1-p) ), μ⇒K␣s
igma⇒L
```

- The Binomial dist graph program will first display μ, then display σ, and then show the graph of the probability distribution. To resume the execution of the program after each of these is displayed, press [COMP] .

Notice that Binomial dist graph also gives you a convenient way of calculating all the binomial probabilities p(x) for x = 0, 1, 2, . . . , n. After the graph is drawn, press [TITLE / DATA] [=] and you will see the probabilities displayed on the right hand side of the screen with B[1 , i + 1] containing p(x = i). To return to the data screen, press

TITLE DATA and 2ndF T-G-D.

- Probabilities are stored in matrix location B before the graph is drawn. If you receive an ERROR 4 message while executing this program, press 2ndF M.CK to check the available memory. If there are no bytes remaining, you may have to delete other stored data or another program so that this program can run.

NORMAL DISTRIBUTION OVERLAY

In the next chapter you will study the normal distribution. The normal distribution has been studied extensively and can in many cases be used to approximate probabilities for discrete random variables. Your statistics text will very likely give you "rules-of-thumb" as to when these approximations are valid. Geometrically, whenever one considers an approximation by the normal distribution, the bell-shaped curve of the normal probability distribution should "fit" nicely over the graph of the distribution it is approximating. The following program, Normal dist overlay, draws a graph of the normal distribution over the probability histogram you have generated by the program Binomial dist graph. The normal random variable is set to have the same mean and standard deviation as the binomial random variable for which you have drawn the probability histogram.

PROGRAMMING

Select the AER-II mode and you will see on the display 1Ø: TITLE?. Type in, using the letters on the right keyboard, SHIFT Normal ␣ dist ␣ overlay and press ENT to store the program name. Input the program at the M: prompt as:

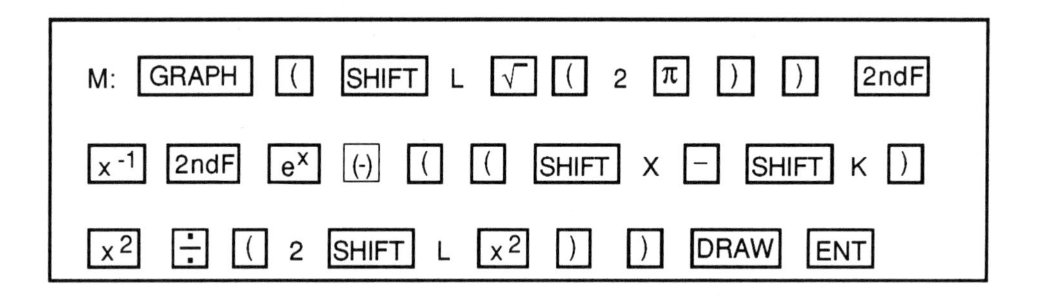

Once the program is entered, your calculator screen should look exactly like the following:

M : GRAPH (L √ (2 π))
-1 eˣ - ((X−K)² ÷ (2 L²
)) DRAW

EXERCISES

1) Use the program **Binomial dist graph** to construct graphs for the binomial distributions with $n = 10$ and $p = 0.10$, $n = 10$ and $p = 0.35$, $n = 10$ and $p = 0.60$, and $n = 10$ and $p = 0.90$. How is the shape of the graph changing as p increases?

2) Use the program **Binomial dist graph** to construct graphs for the binomial distributions with $n = 5$ and $p = 0.20$ and with $n = 15$ and $p = 0.56$. In each case, overlay the graph of the normal distribution with the program **Normal dist overlay**. Which bell shaped curve best fits (in terms of the areas under the two being nearly the same) the underlying binomial distribution?

THE POISSON DISTRIBUTION

Let's continue our discussion of programming on the EL-5200 with a further look at the AER-II mode. Expressions in this mode may use lowercase letters, Greek letters, numeric characters reduced in size and other special symbols in addition to uppercase letters. Lowercase letter variables have their assigned values stored in the program memory whereas the uppercase variables values are stored in the store memories. Lowercase variables may consist of a single letter or a combination of letters (with no separation marks). In the AER-II mode, the first time a program finds a lowercase variable, say c, the calculator will automatically prompt for the value with "c=?".

To use an uppercase variable, say D, in a program written in the AER-II mode, you must precede the variable with the SHIFT key in order to capitalize it. If you do not wish to input the value of the uppercase variable into the appropriate store memory with the STO key, you can use f()=/? in the program following the variable name to result in the "D = ?" prompt when the program is executed.

Another discrete distribution is the Poisson probability distribution. The random variable x counts the number of successes (events) that occur in a specific period of time, area, or volume and λ is the average (mean) number of events that occur during the given time period, or over the specific area or volume. The other characteristics of a Poisson random variable are:

- The probability that an event occurs in any specified time period or area or volume is the same for all units, and
- The occurrences of the events are independent.

The probability distribution for a Poisson random variable x is given by the formula

$$p(x) = \frac{\lambda^{x} e^{-\lambda}}{x!} \quad \text{for } x = 0, 1, 2, 3, \ldots$$

PROGRAMMING

Place the calculator in the AER-II mode. When the prompt 11: TITLE? appears on the screen, press SHIFT P o i s s o n and press ENT to store the program name. Input the program for the Poisson distribution as follows:

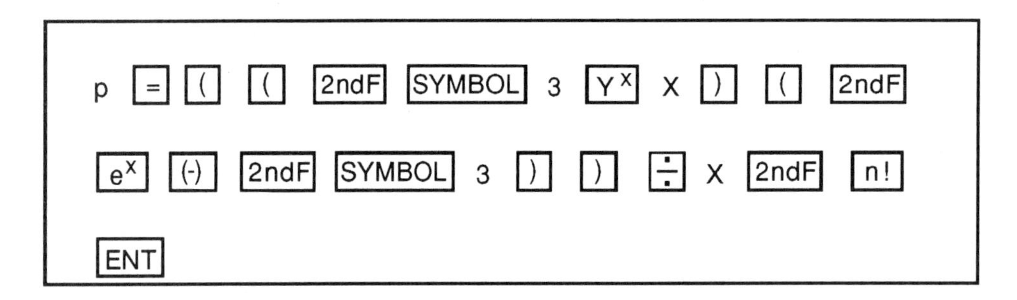

Once the program is entered, your calculator screen should look exactly like the following:

```
M: p = ( ( Y Yx x ) ( ex -Y
)) ÷ x!
```

- Note: The symbol λ does not appear in the special character selection so the symbol Y was used in the calculator program to represent the average (mean) number of events that occur during the given time period.

EXERCISES	ANSWERS
1) For the Poisson random variable x with $\lambda = 6.2$, find $P(x = 8)$.	1) .1099
2) For the Poisson random variable x with $\lambda = 3.86$, find $P(x \geq 1)$.	2) 1 – .0211 = .9789
3) A fisherman catches an average of 15.3 fish per day in the Jordan River. Find the probability that on any given day the fisherman catches 20 fish.	3) .0460

GRAPH OF THE POISSON DISTRIBUTION

The Poisson probability distribution varies with different values of λ, the expected number of successes per unit. The following program will show the Poisson distribution probability histogram for the number of successes, x, between 0 and INT(2λ) + 1. This program stores the Poisson probabilities for $0 \le x \le \text{INT}(2\lambda) + 1$ in matrix location G[1] and uses the LINE function to construct the histogram in a similar manner to the program Binomial dist graph. It does not use the histogram function G(HI) in the STAT mode for that will graph only frequency distributions, not probability distributions.

The following graphs are examples of those produced using the program Poisson dist graph.

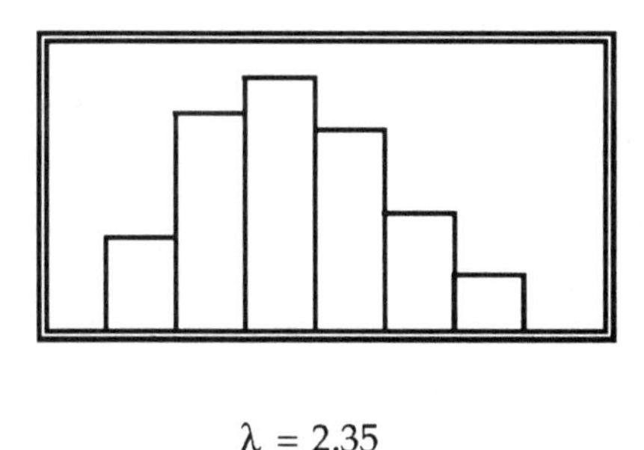

$\lambda = 2.35$

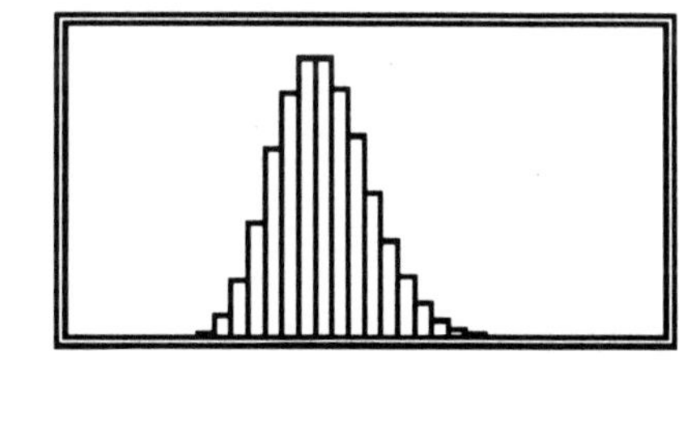

$\lambda = 8$

PROGRAMMING

Select the AER-II mode and you will see on the display 12: TITLE?. Type in, using the letters on the right keyboard, SHIFT P o i s s o n ⊔ d i s t ⊔ g r a p h press ENT to store the program name. Input the program at the M: prompt as:

---

[1]Matrix dimensions in your calculator must be whole numbers. If 2λ is not a whole number, an error message will result when the program Poisson specifies the dimension of the matrix.

2ndF INT (2 2ndF SYMBOL 3) + 1 STO SHIFT

N 2ndF DIM SHIFT G 2ndF [SHIFT N + 1 2ndF]

Ø STO SHIFT X 2ndF ↳ 2ndF 1 (SHIFT N +

1) 2ndF > SHIFT X 2ndF ≡Y →[ ] 2ndF ⌝ 2ndF

≡Y →[ ] Ø STO SHIFT H ↳ SHIFT H + 1 STO SHIFT

H SHIFT G 2ndF [SHIFT H 2ndF] 2ndF > SHIFT

G 2ndF [SHIFT H + 1 2ndF] 2ndF ≡N →[ ]

2ndF ⌝ 2ndF ≡N →[ ] 2ndF 4 2ndF 2 Ø STO

SHIFT C 2ndF ↳ 2ndF 3 SHIFT C 2ndF ≠ (

SHIFT N + 1 2ndF ≡Y →[ ] 2ndF ⌝ 2ndF ≡Y →[ ]

2ndF SUB *(at the* 1 *: prompt, enter the following)* 2ndF SYMBOL

3 Y^X SHIFT X (2ndF e^X (-) 2ndF SYMBOL 3)

÷ SHIFT X n! STO SHIFT G 2ndF [SHIFT X +

1 2ndF] SHIFT X + 1 STO SHIFT X 2ndF SUB

(at the 2 *: prompt, enter the following)* RANGE (-) 1 , SHIFT

Note: The program Poisson dist graph is continued on the next page.

N [+] 1 [,] 1 [,] Ø [,] [SHIFT] G [2ndF] [[] [SHIFT] H

[2ndF] []] [,] [SHIFT] G [2ndF] [[] [SHIFT] H [2ndF] []]

[÷] 5 [2ndF] [SUB] *(at the* [3] *: prompt, enter the following)* [2ndF]

[SYMBOL] 3 [Y^X] [SHIFT] C [(] [2ndF] [e^X] [(-)] [2ndF]

[SYMBOL] 3 [)] [÷] [SHIFT] C [n!] [STO] [SHIFT] P [SHIFT]

C [−] .5 [STO] [SHIFT] A [SHIFT] C [+] .5 [STO] [SHIFT]

B [2ndF] [LINE] [SHIFT] A [,] [SHIFT] P [,] [SHIFT] B [,]

[SHIFT] P [DRAW] [2ndF] [LINE] [SHIFT] A [,] [SHIFT] P [,]

[SHIFT] A [,] Ø [DRAW] [2ndF] [LINE] [SHIFT] B [,] [SHIFT]

P [,] [SHIFT] B [,] Ø [DRAW] [SHIFT] C [+] 1 [STO] [SHIFT]

C [2ndF] [SUB] *(at the* [4] *: prompt, enter the following)* [2ndF]

[SYMBOL] 6 [=] [2ndF] [SYMBOL] 3 [,] s i g m a [=] [√]

[2ndF] [SYMBOL] 3 [,] [2ndF] [SYMBOL] 6 [STO] [SHIFT] K

[␣] s i g m a [STO] [SHIFT] L [ENT]

Once the program is entered, your calculator screen should look exactly like the following. Press [2ndF] [▽] to see the contents of the subroutines.

```
M: INT (2γ) +1⇒N DI
M G[N+1] ∅⇒X ↳ [1] (N+
1)>X ≡Y→[ ↰ ] ∅⇒H ↳ H+
1⇒H G[H]>G[H+1] ≡N→
[ ↰ ] [4] [2] ∅⇒C ↳ [3] C≠(N+
1) ≡Y→[ ↰ ]
```

```
[3] : γ Y^x C (e^x - γ) ÷ C ! ⇒
P C - .5⇒ A C + .5⇒ B LIN
E A , P, B , P DRAW LIN
E A , P, A , ∅ DRAW LIN
E B , P , B, ∅ DRAW C +1
⇒ C
```

```
[1] : γ Y^x X (e^x - γ) ÷ X ! ⇒
G[X+1] X+1 ⇒X
```

```
[4] : μ = γ, sigma = √γ, μ
⇒K ⊔ sigma ⇒L
```

```
[2] : RANGE -1, N +1, 1
, ∅, G[H] ,G[H] ÷ 5
```

- The **Poisson dist graph** program will first display μ, then display σ, and then show the graph of the probability distribution. To resume the execution of the program after each of these is displayed, press [COMP] .

- If you wish to compare distributions generated by several programs, you can name the unprotected storage locations. For instance, the program **Binomial dist graph** stores data in matrix location B. To name this matrix location, press [TITLE / DATA] and with the blinking cursor over the letter B, press [▷]. Type in, using the letters on the right keyboard the name BINOMIAL. Press [=] to store the name. The matrix location must have data in it before the calculator will accept a name for the location. You can assign the name POISSON to matrix location G by following this same procedure.

Notice that Poisson dist graph also gives you a convenient way of calculating all the binomial probabilities p(x) for x = 0, 1, 2, . . . , n. After the graph is drawn, press

TITLE/DATA [=] and you will see the probabilities displayed on the right hand side of the screen with G[1 , i + 1] containing p(x = i). To return to the data screen, press

TITLE/DATA and [2ndF] [T-G-D].

EXERCISES

1) Use the program Poisson dist graph to construct graphs for the Poisson distributions with $\Upsilon = 1.75$, $\Upsilon = 5$ and $\Upsilon = 10$. How is the shape of the graph changing as Υ increases?

2) Use the program Poisson dist graph to construct graphs for the Poisson distributions with $\Upsilon = 2.63$ and $\Upsilon = 8$. In each case, overlay the graph of the normal distribution with the program Normal dist overlay. Which bell-shaped curve best fits (in terms of the areas under the two being nearly the same) the underlying Poisson distribution?

THE HYPERGEOMETRIC DISTRIBUTION

The hypergeometric random variable has a discrete probability distribution whose characteristics are:

- Each draw or trial results in one of two outcomes, success or failure.
- The experiment consists of drawing at random n elements without replacement from a set of N elements, r of which are called success and N − r of which are called failure.

The probability distribution for the hypergeometric random variable is given by the formula

$$p(x) = \frac{\binom{r}{x}\binom{N-r}{n-x}}{\binom{N}{n}} \quad \text{for } x = \text{maximum}(0, n-(N-r)), \ldots, \text{minimum}(r, n)$$

PROGRAMMING

Place the calculator in the AER-II mode. When the prompt 13: TITLE? appears on the screen, press [SHIFT] H y p g e o m and press [ENT] to store the program name. Input the program for the hypergeometric distribution as follows:

[SHIFT] N [f()=/?] [⊔] p [=] [(] [(] r [2ndF] [nCr] x [)]
[(] [(] [SHIFT] N [−] r [)] [2ndF] [nCr] [(] n [−] x [)]
[)] [)] [÷] [(] [SHIFT] N [2ndF] [nCr] n [)] [ENT]

- The command [SHIFT] N [f()=/?] defines N as a variable in the AER-II mode and prompts for the value of N when the program is run in the COMP mode.

Once the program is entered, your calculator screen should look exactly like:

```
M: N=?␣p=((rCx)((
N-r)C(n-x)))÷(NC
n)
```

• After the initial values of N, n, r, and x are entered into this program, you may wish to compute again with some of the same values. Simply press [COMP] and the value of the variable that was previously used will be again entered for the value of that variable. If you are not sure which value is stored in the calculator, press [PB] and the value held in memory will be displayed. Press [COMP] to enter this value or the proper numerical key to change the value.

• Now that you are putting more programs in your calculator, you may pass a particular program when searching. If so, use the second function key and the program key ([2ndF] [PRO] (TITLE)) to scroll backwards through the programs.

EXERCISES	ANSWERS
1) For a hypergeometric experiment with N = 8, r = 3, and n = 5, find P(x = 2).	1) .5357
2) A committee of size 10 is to be selected from a group of 15 men and 12 women. Find the probability that there will be exactly 5 women on the committee.	2) .2819
3) A box contains 5 green balls, 6 blue balls, 4 white balls, and 5 red balls. All balls are identical except for color. Four balls are chosen at random from the box. Find the probability that exactly three balls are red or all four balls are blue.	3) .0310+.0031 = .0341

GRAPH OF THE HYPERGEOMETRIC DISTRIBUTION

The hypergeometric probability distribution varies with different values of N, the total number of items under consideration, n, the lot size, and r, the total number of items called success. The following program stores the hypergeometric probabilities in matrix location J and uses the LINE function to construct the probability histogram. You have probably noticed that the hypergeometric distribution is not defined for all values of x between 0 and n since $\binom{n}{x}$ only makes sense in an applied problem when $x \leq n$. Thus, in the following program, as in the definition of hypergeometric probabilities, x ranges from the maximum of 0 and $n - (N - r)$ to the minimum of r and n. The following graphs are examples of those produced using the program Hypgeom dist graph.

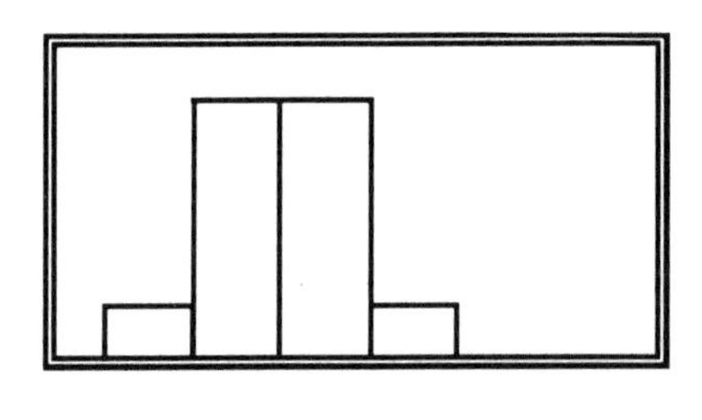

N = 10, n = 5, r = 3

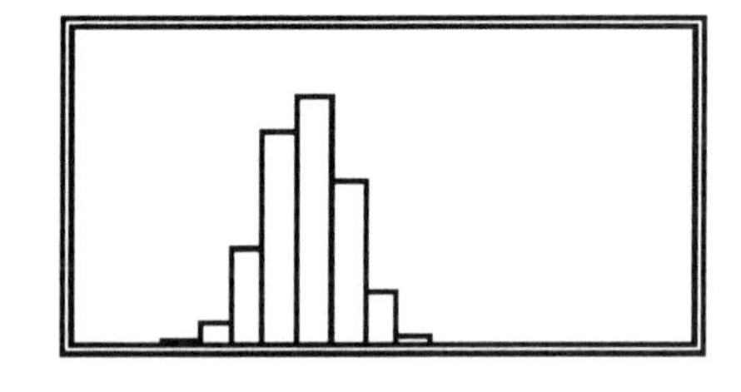

N = 26, n = 15, r = 10

PROGRAMMING

Select the AER-II mode and you will see on the display 14: TITLE?. Type in [SHIFT] H y p g e o m [␣] d i s t [␣] g r a p h and press [ENT] to store the program name. Input the program at the M: prompt as:

```
2ndF DIM SHIFT J 2ndF [ n + 1 2ndF ] n STO
SHIFT R SHIFT N f()=/? ⊔ ( SHIFT R − (
SHIFT N − r ) ) STO SHIFT T ⊔ Ø 2ndF >
SHIFT T 2ndF ≡Y→[ ] Ø STO SHIFT X 2ndF ≡Y→[ ]
2ndF ≡N→[ ] SHIFT T STO SHIFT X 2ndF ≡N→[ ]
2ndF ↳ 2ndF 1 SHIFT R 2ndF > r 2ndF ≡Y→[ ]
r STO SHIFT M 2ndF ≡Y→[ ] 2ndF ≡N→[ ] SHIFT R
STO SHIFT M 2ndF ≡N→[ ] ( SHIFT M + 1 )
2ndF > SHIFT X 2ndF ≡Y→[ ] 2ndF ↰ 2ndF ≡Y→[ ]
Ø STO SHIFT H 2ndF ↳ SHIFT H + 1 STO SHIFT
H SHIFT J 2ndF [ SHIFT H 2ndF ] 2ndF >
SHIFT J 2ndF [ SHIFT H + 1 2ndF ] 2ndF
≡N→[ ] 2ndF ↰ 2ndF ≡N→[ ] 2ndF 4 2ndF 2 Ø
2ndF > SHIFT T 2ndF ≡Y→[ ] Ø STO SHIFT C 2ndF
≡Y→[ ] 2ndF ≡N→[ ] SHIFT T STO SHIFT C 2ndF
```

Note: The program Hypgeom dist graph is continued on the next page.

≡N→[ ] 2ndF ↳ 2ndF 3 SHIFT R 2ndF > r 2ndF

≡Y→[ ] r STO SHIFT M 2ndF ≡Y→[ ] 2ndF ≡N→[ ]

SHIFT R STO SHIFT M 2ndF ≡N→[ ] SHIFT C 2ndF

≠ (SHIFT M + 1) 2ndF ≡Y→[ ] 2ndF ↰

2ndF ≡Y→[ ] 2ndF SUB *(at the* 1*: prompt, enter the following)*

((r 2ndF nCr SHIFT X) ((SHIFT N −

r) 2ndF nCr (SHIFT R − SHIFT X)))

÷ SHIFT N 2ndF nCr SHIFT R STO SHIFT J 2ndF

[SHIFT X + 1 2ndF] SHIFT X + 1 STO

SHIFT X 2ndF SUB *(at the* 2 *: prompt, enter the following)*

RANGE (-) 1 , SHIFT R + 1 , 1 , Ø , SHIFT

J 2ndF [SHIFT H 2ndF] , SHIFT J 2ndF [

SHIFT H 2ndF] ÷ 5 2ndF SUB *(at the* 3 *: prompt,*

enter the following) ((r 2ndF nCr SHIFT C) ((

Note: The program Hypgeom dist graph is continued on the next page.

[SHIFT] N [-] r [)] [2ndF] [nCr] [(] n [-] [SHIFT] C [)] [)] [)] [÷] [SHIFT] N [2ndF] [nCr] n [STO] [SHIFT] P [SHIFT] C [-] .5 [STO] [SHIFT] A [SHIFT] C [+] .5 [STO] [SHIFT] B [2ndF] [LINE] [SHIFT] A [,] [SHIFT] P [,] [SHIFT] B [,] [SHIFT] P [DRAW] [2ndF] [LINE] [SHIFT] A [,] [SHIFT] P [,] [SHIFT] A [,] Ø [DRAW] [2ndF] [LINE] [SHIFT] B [,] [SHIFT] P [,] [SHIFT] B [,] Ø [DRAW] [SHIFT] C [+] 1 [STO] [SHIFT] C [2ndF] [SUB] *(at the* [4] *: prompt, enter the following)* [2ndF] [SYMBOL] 6 [=] n [×] r [÷] [SHIFT] N [,] sigma [=] [√] [(] [(] r [(] [SHIFT] N [-] r [)] n [(] [SHIFT] N [-] n [)] [)] [÷] [(] [SHIFT] N [x^2] [(] [SHIFT] N [-] 1 [)] [)] [)] [,] [2ndF] [SYMBOL] 6 [STO] [SHIFT] K [⊔] sigma [STO] [SHIFT] L [ENT]

Once the program is entered, your calculator screen should look exactly like the following. Recall that you press 2ndF ▽ to view the contents of the subroutines.

```
M: DIM  J[n+1 ] n⇒R N
= ?⊔(R-(N-r))⇒T⊔ Ø
> T ≡Y→[ Ø⇒ X ] ≡N→[ T⇒
X] ↳ [1] R > r ≡Y→[ r⇒ M ] ≡
N→[ R⇒ M ] ( M +1 )> X ≡Y
→[ ↰ ]Ø⇒H ↳ [H+1]⇒HJ [H
]> J [H+1] ≡N→[ ↰ ] [4] [2]
Ø > T ≡Y→[ Ø⇒ C] ≡N→[ T
⇒C] ↳ [3] R > r ≡Y→[ r⇒ M ]
≡N→[ R⇒ M ] C ≠ ( M +1 ) ≡
Y→[ ↰ ]
```

```
[1] : ( (rCX ( (N - r)C (
R - X) ) )÷(N C R)⇒J[X
+1] X+1⇒X
```

```
[2] : RANGE  -1 , R +1 , 1
, Ø, J[H] ,J[H] ÷ 5
[3] : ( (rCC ( (N - r)C (
n-C))) ÷ (N C n)⇒P C-
.5 ⇒A C + .5 ⇒B LINE A
, P, B , P DRAW LINE A
, P, A , Ø DRAW LINE B
, P, B , Ø DRAW C +1⇒C
```

```
[4] :μ=n x r÷N,sigma=
√( ( r ( N - r ) n (N-n))
÷ (N² (N -1))), μ⇒K⊔
sigma⇒L
```

- The Hypgeom dist graph program will first display μ, then display σ, and then show the graph of the probability distribution. To resume the execution of the program after each of these is displayed, press COMP .

As in the previous programs, Hypgeom dist graph also gives you a convenient way of calculating all the hypergeometric probabilities p(x). After the graph is drawn, press [TITLE / DATA] [=] and you will see the probabilities displayed on the right hand side of the screen with J [1 , i + 1] containing p(x = i). To return to the data screen, press [TITLE / DATA] and [2ndF] [T-G-D].

EXERCISES

1) Use the program Hypgeom dist graph to construct a graph for the hypergeometric distributions with N = 18, n = 7 and r = 5. Look at the scale on the x axis (each tic mark represents one unit) and the J data matrix to see for which values the probability has been calculated. Explain what you see.

2) Use the program Hypgeom dist graph to construct graphs for the hypergeometric distributions with N = 10, n = 6 and r = 3 and with N = 20, n = 8 and r = 6. In each case, overlay the graph of the normal distribution with the program. Which bell-shaped curve best fits (in terms of the areas under the two being nearly the same) the underlying hypergeometric distribution?

3) a) Construct a graph of the binomial distribution for n = 50 and p = .1. The Sharp EL-5200 has no mechanism to store graphs, so record the graph on paper while you notice its skewness, mean and standard deviation. Remember that the binomial probabilities are stored in matrix location B.

   b) Construct a graph of the Poisson distribution for $\lambda = 5$. (Remember that the Poisson probabilities are store in matrix location G.) Does it appear that the graph of the Poisson distribution would "fit" over the graph of the binomial distribution you constructed in a)? What is the relation of the mean of the binomial to the mean of the Poisson in this problem?

   c) If you have the available memory in your calculator, repeat parts a) and b) with n = 100.

4) a) Construct a graph of the hypergeometric distribution with $N = 50$, $n = 10$ and $r = 5$. Record the graph on paper while you notice its skewness, mean and standard deviation. Also record the probability that $x = 2$.

b) Construct a graph of the binomial distribution for $n = 10$ and $p = .10$. Compare the skewness, mean and standard deviation to the graph you drew in part a). What is the probability that $x = 2$?

c) Notice that the $p = .10$ for the binomial equals r/N for the hypergeometric. Do you think the binomial distribution with $p = r/N$ could in some cases be used to approximate the probabilities for the hypergeometric distribution? If so, when?

- If you get a ERROR 2 message when trying this problem with other values, it is because your calculator cannot compute $\binom{n}{x}$ when $n \geq 70$.

CHAPTER 5

THE NORMAL DISTRIBUTION

There are experiments whose sample spaces contain other than a countable number of simple events. Continuous random variables can assume values within an interval or intervals of real numbers. Heights, weights, temperatures, the time required to perform a certain task, and the length of time between arrivals of trucks at an unloading dock are all examples of continuous random variables.

The function f(x) is called a probability density (distribution) function of the continuous random variable x if

$$f(x) \geq 0 \quad \text{for} \quad -\infty < x < \infty \quad \text{and} \quad \int_{-\infty}^{\infty} f(x)\, dx = 1 .$$

The probability that a continuous random variable x will fall in a certain interval [a,b] equals the area under the graph of the probability density function between the two values a and b. The probability of any specific value of a continuous random variable is zero since single numeric values have no width. Probabilities are calculated for intervals of real values. Thus, for the continuous random variable x, $P(a \leq x \leq b) = P(a \leq x < b) = P(a < x \leq b) = P(a < x < b)$.

A continuous random variable known as the normal random variable plays an important role in statistical inference. The equation of the normal density function is

$$f(x) = \frac{1}{\sigma\sqrt{2\Pi}}\, e^{-\frac{(x-\mu)^2}{2\sigma^2}} \qquad \text{for} \quad -\infty < x < \infty .$$

The normal distribution is symmetric about its mean μ and its spread is determined by the standard deviation σ. The graph of the normal density function with mean μ and variance σ^2 is

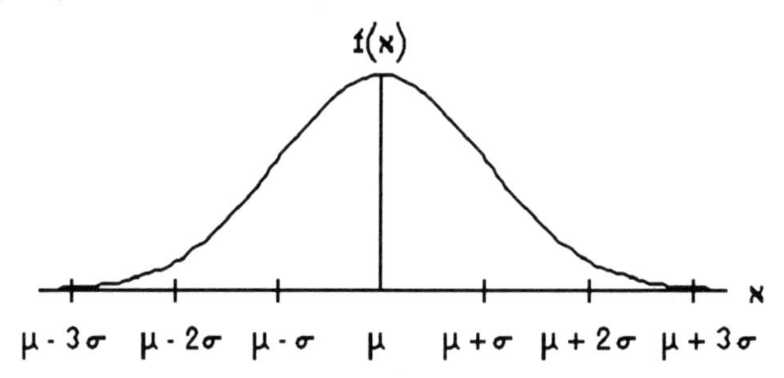

Actually, the normal probability distribution is a family of bell-shaped curves with each particular curve differentiated from the others by its mean μ and standard deviation σ. Since the equation of the normal density function is tedious to type, enter it as a program and there will be fewer keystrokes needed to draw its graph.

PROGRAMMING

Select the AER-II mode. You will see on the display 15: TITLE? Type in, using the letters on the right keyboard, [SHIFT] N o r m a l [⊔] g r a p h s and press [ENT] to store the program name. Input the program at the M: prompt as follows:

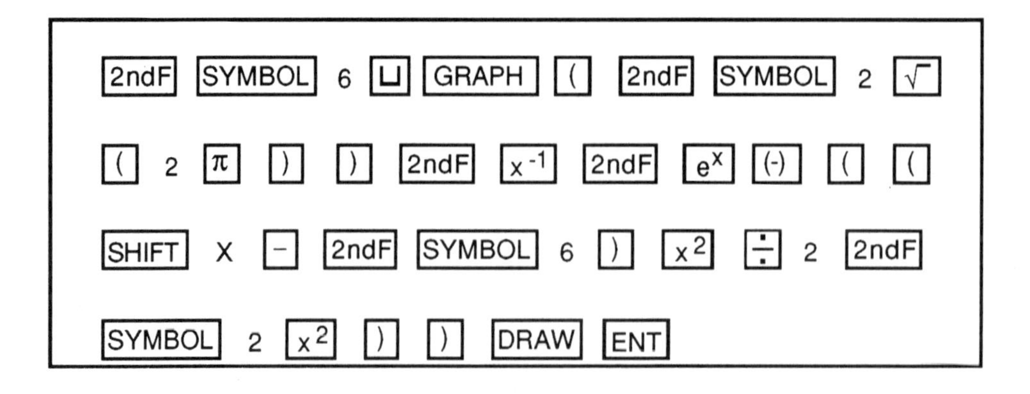

- Note that the program calls the standard deviation using the symbol β since there is no symbol for σ listed in the available symbol characters. If you prefer not to use this symbol, replace β in the above program by the letters s i g m a.

Once the program is entered, your calculator screen should look exactly like the following:

```
M: μ␣ GRAPH  (β√(2π
))⁻¹ eˣ − ((X − μ)² ÷ (2
β²)) DRAW
```

To draw an accurate graph of a particular normal density function, you must know the values of μ and σ. To see the effect on the graph with changing values of μ and σ, let's graph the normal distribution for $\mu = 2$ and $\sigma = 5$ and then overlay a graph of the normal distribution for $\mu = 4$ and $\sigma = 3$.

Before executing the above program, return to the COMP mode, press RANGE and set the following values in order to obtain a "good" view of both graphs:

Xmin = - 8, Xmax = 8, Xscl = 1, Ymin = 0, Ymax = . 2, Yscl = . 05

Press COMP and enter the values of $\mu = 2$ and σ (β) = 5 when requested. When the graph appears on the screen, press COMP again and enter $\mu = 4$ and σ (β) = 3. It will take the Sharp EL-5200 a while to draw this graph. Why don't you draw by hand a rough sketch of what you expect the graph to look like while you are waiting?

Once the graphs appear, your graphics screen display should look like

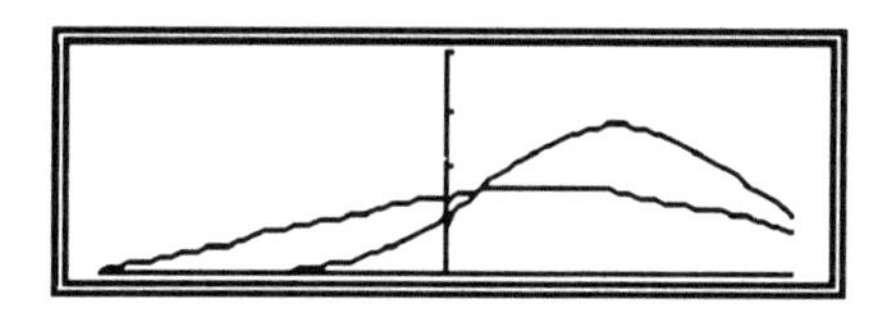

Notice from your graphs that the highest point of each normal curve occurs at the mean μ. Larger values of μ shift the graph to the right and smaller values of μ shift the graph to the left. Since the standard deviation determines the width (spread) of the curve and the total area under any normal curve equals one, larger values σ of result in wider, flatter curves and smaller values of σ give taller, narrower curves.

STANDARD NORMAL DISTRIBUTION

It is easier to compare normal distributions having different values of μ and σ if these distributions are transformed to a common form called the standard normal distribution. The standard normal variable, denoted by z, has a mean of zero and a standard deviation equal to one. The equation of the density function for the standard normal distribution is $f(z) = \frac{1}{\sqrt{2\Pi}} e^{-\frac{1}{2}z^2}$ for $-\infty < z < \infty$ with graph

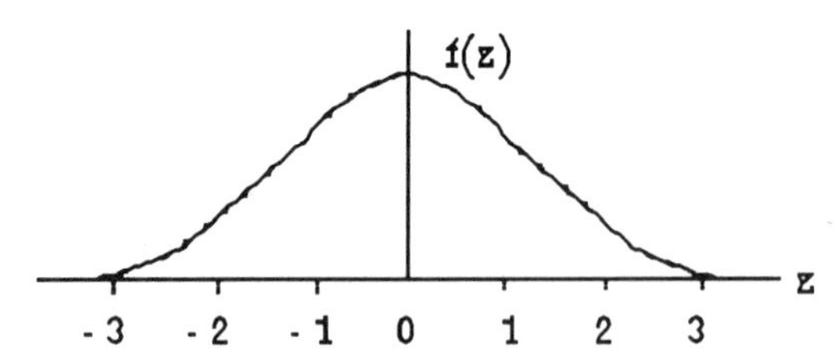

If you have not used your graphics screen since the normal graphs for $\mu = 2, \sigma = 5$ and $\mu = 4, \sigma = 3$ were drawn, you can overlay the graph of the standard normal distribution on these two graphs by executing the program Normal graphs using the values $\mu = 0$ and $\sigma = 1$. Compare the centers and spread of values in the three graphs.

Next, reset the range parameters Xmin and Xmax to -5 and 5 respectively. To what value should Ymax be set in order to see the "top" of the standard normal distribution on your graph? (Hint: Substitute z = 0 in the equation of the standard normal density function and calculate f(0) with the keystrokes [( ] [√] [( ] 2 [π] [ )] [ )] [x^{-1}] .) Set Ymax to this value. Ymin should be set at 0. Draw the graph of the standard normal density function using the program Normal graphs with $\mu = 0$ and $\sigma = 1$. Once again enter the range parameters and reset Xmin and Xmax to -3 and 3. Draw the graph of the standard normal distribution. Explore with different values of Xmin and Xmax to see which values give the "best-looking" graph. You will notice that even though the standard normal density function, or z distribution, has domain of all real numbers, practically all of the graph appears between -3 and 3.

If you have a knowledge of calculus, you will realize that areas under the standard normal density function cannot be obtained by standard integration methods. However, these areas can be found to any desired degree of accuracy by approximation techniques. Tables of areas under the standard normal curve are found in most statistics texts. The following program calculates the area under the standard normal curve between the mean, z = 0, and a specific positive value of z by a procedure known as Hasting's best approximation.

PROGRAMMING

Select the AER-II mode. You will see on the display 16: TITLE? Type in, using the letters on the right keyboard, [SHIFT] Normal [␣] prob and press [ENT] to store the program name. Input the program at the M: prompt as follows:

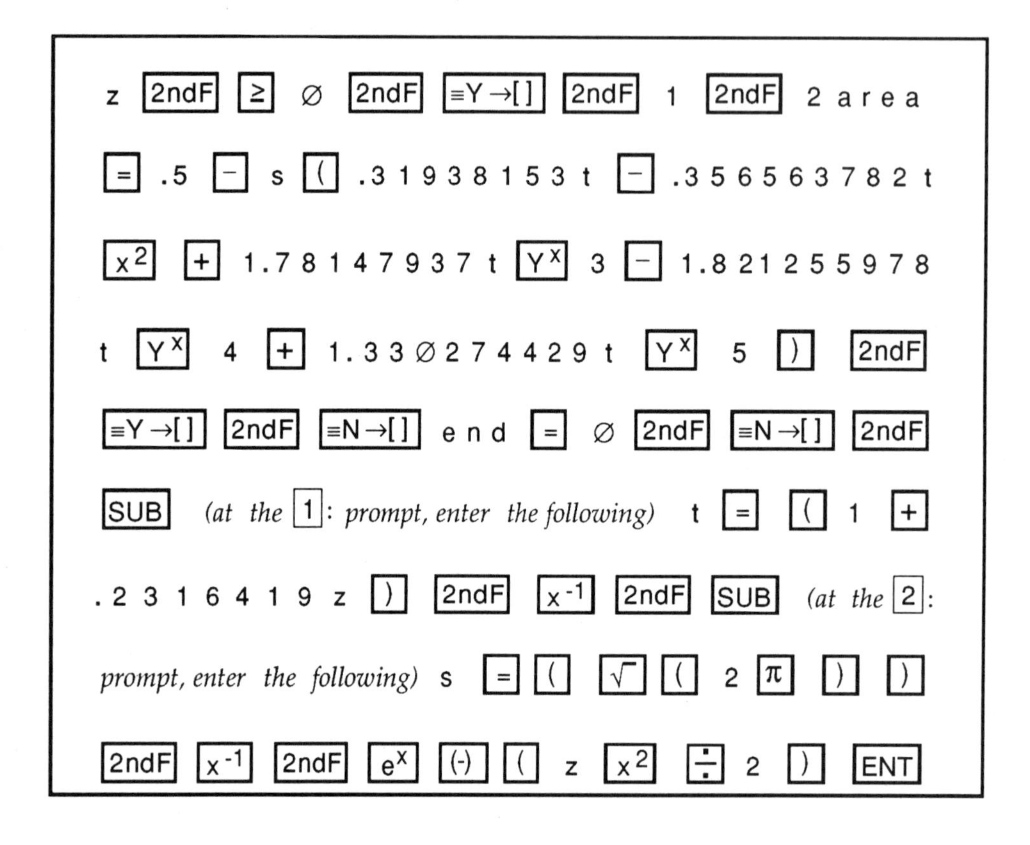

Once the program is entered, your calculator screen should look exactly like the following:

```
M: z>= Ø ≡Y→[ 1 2 area
=.5 – s (.31938153 t
–.356563782 t² + 1.
78147937 t Yˣ 3 – 1.8
21255978 t Yˣ 4 + 1.3
3Ø274429 t Yˣ 5) ] ≡N
→[end=Ø]
```

Press 2ndF ▽ to see the contents of subroutine 1 which should appear as

```
1: t=(1+.2316419z
 )^-1
```

Press 2ndF ▽ to see the contents of subroutine 2 which should appear as

```
2: s=(√(2π))^-1 e^x - (
z^2 ÷ 2)
```

Place the calculator in the COMP mode and enter the value of z when requested. The program returns the area between 0 and the entered value of z whenever $z \geq 0$. For example, to find $P(0 < z < 1.53)$, the area between 0 and $z = 1.53$, enter 1.53 at the z=? prompt and "area = .4370" is returned.

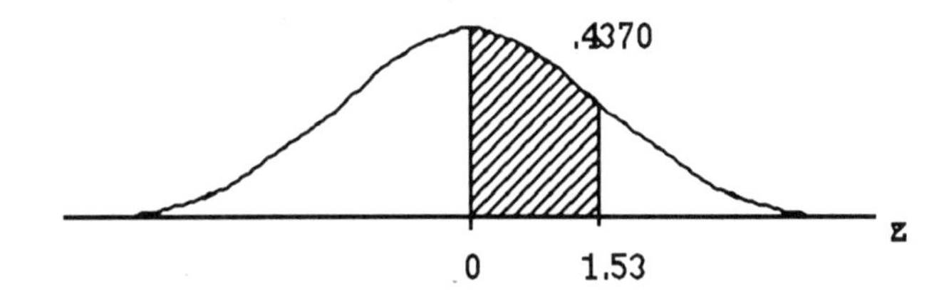

- Note: The above program is designed to give the area under the standard normal density function between 0 and the entered value of z for $z \geq 0$. The program is *not* designed to accurately give areas for negative values of z. Thus, if a negative value of z is entered, the message "end = 0" is returned to remind you that you have entered an incorrect value of z. The following information will allow you to compute areas between 0 and any real value of z.

Since the normal curve is symmetric about its mean, half of the area lies to the left of the mean of zero and half to the right of zero.

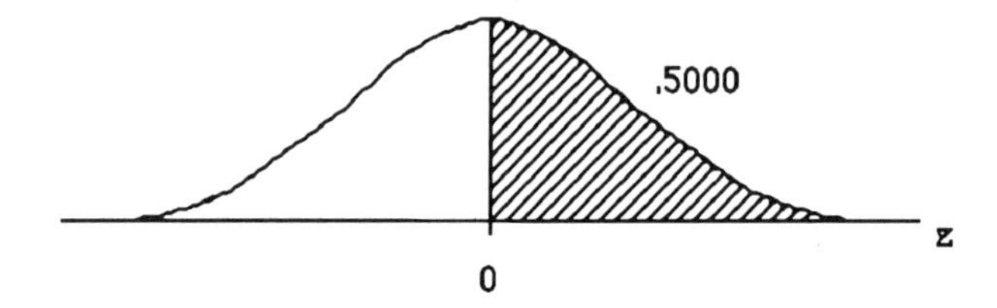

Areas between zero and negative values of z are obtained using the symmetry of the density function rather than using the program Normal prob. Other related areas can easily be calculated by using the area(s) obtained from the program and a reference sketch of the standard normal distribution.

For example, to find P(z < -1.67) , enter 1.67 at the z = ? prompt and obtain the answer "area = .4525". Recall that .4525 is the area between 0 and 1.67, which by symmetry, also equals the area between 0 and -1.67. From the sketch below, you see that the area to the left of -1.67 equals .5000 – .4525 = .0475.

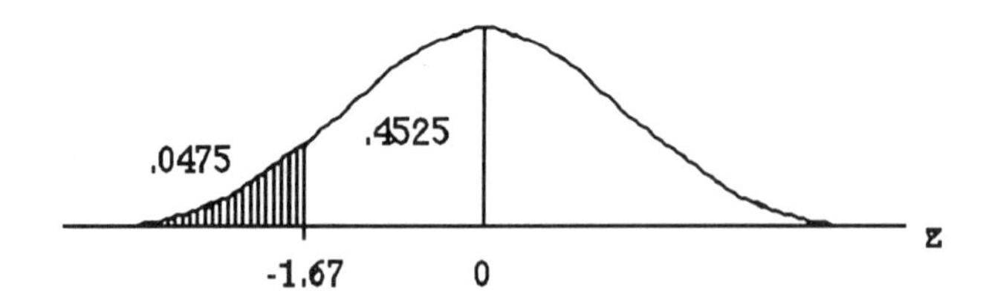

Let's look at another example. To find P(.55 ≤ z < 1.64), enter 1.64 at the z = ? prompt and obtain "area = .4495". Store this value in memory using the [⇒M] key. Execute the program again and enter .55 at the z = ? prompt and obtain the value area = .2088. Press [2ndF] [M+] (which subtracts the value on the screen from memory) and press RM to obtain the desired area of .2407. (Note that the same procedure would be used to find the area between -.55 and -1.64.)

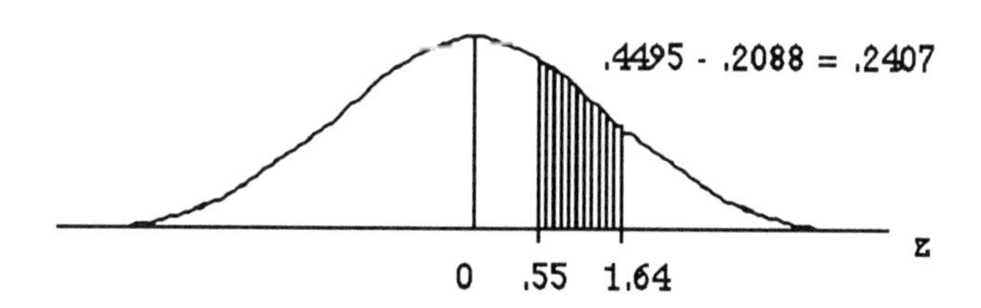

To find probabilities associated with normally distributed variables other than z, say x, first convert x to z using the formula $z = \frac{x - \mu}{\sigma}$. The above procedures can then be used to find the desired probability.

There will be times when you wish to use the above process in reverse. Let's define z_α to be the specific value of the standard normal random variable z such that $P(z > z_\alpha) = \alpha$. Notice that the *probability (area) is the subscript* on the z value. This area, α, is called a *tail area* because it is in the outer portion of the standard normal distribution rather than being an area near the center of the distribution. Graphically, this statement tells us

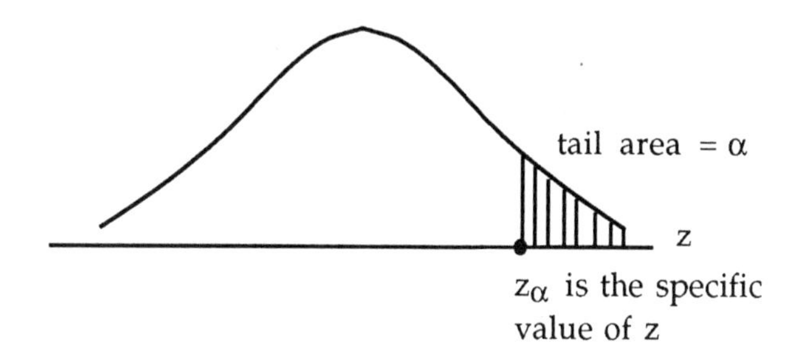

The following program, Normal value, will give the value of z, say z_o , that has a user-specified tail area to the *right* of the point z_o.

PROGRAMMING

Select the AER-II mode. You will see on the display 17: TITLE? Type in, using the letters on the right keyboard, [SHIFT] Normal [⊔] value and press [ENT] to store the program name. Input the program at the M: prompt as follows:

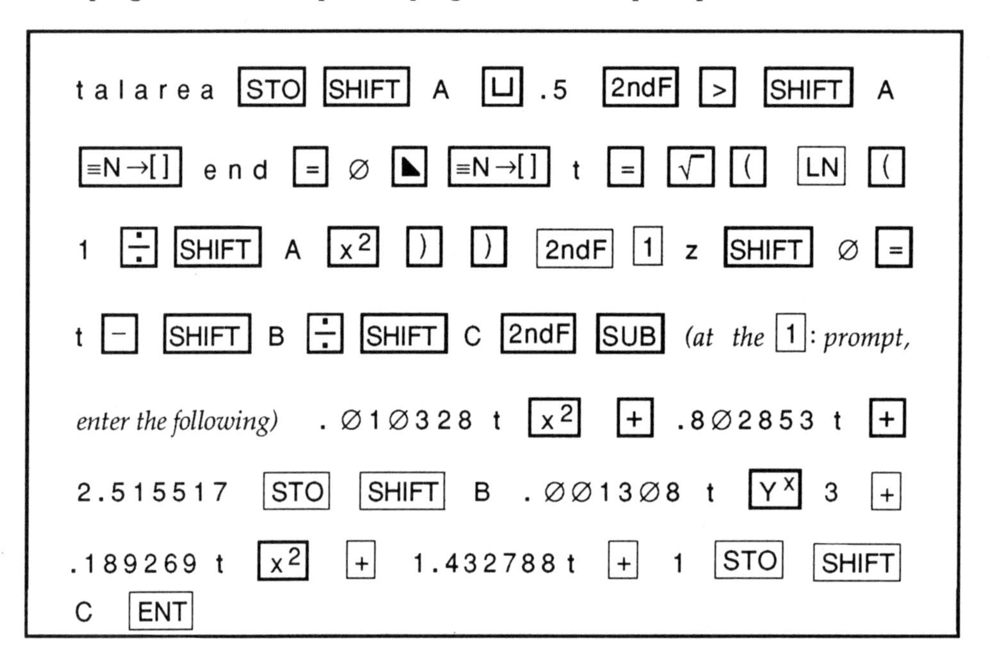

Once the program is entered, your calculator screen should look exactly like the following:

```
M: talarea⇒A ⊔ .5 > A
≡N→[end = Ø ◣] t = √( L
N (1 ÷ A2) ) [1] z0 = t − B
÷ C
```

Press [2ndF] [▽] to see the contents of subroutine [1] which should appear as:

```
[1]: .Ø1Ø328 t² + .8Ø2
853 t + 2.515517⇒B .
ØØ13Ø8 t Yˣ 3 + .1892
69 t² + 1.432788t + 1
⇒C
```

- Note that this program will give the message "end = 0 " whenever an area (probability) of more than .5 is input because the programming design requires the tail area to be less than .5.

As an example, suppose you are asked to find $z_{.1123}$. You are looking for the specific value of z, say z_0, such that .1123 is the area to the right of that value of z. Since the area to the right of the mean of z (recall that the mean of z is 0) is 0.5, this point z_0 must be positive (since the area to the right of any value of $z > 0$ must be less than 0.5). Graphically, we have

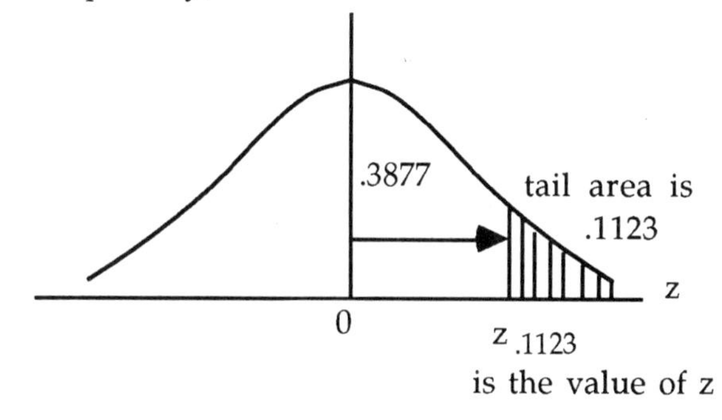

Execute the program Normal value and respond to the message talarea = ? by entering the value .1123, press COMP and you will see "z_0 = 1.2145". This value may be obtained to fewer or more decimal places by pressing TAB followed by the number of decimal places you wish shown.

Now suppose you are asked to find the value of z, say z_o , such that $P(z > z_o) = .8877$. You can still use the program Normal value if you will convert this to a tail area problem. Graphically, we have

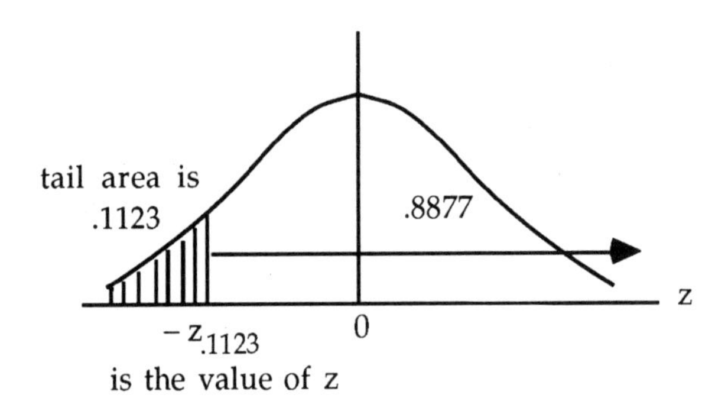

Notice that z_o must be a negative number since the area to the right of 0 is .5 and .8877 > .5. By the symmetry of the standard normal distribution, this problem is exactly the same as the previous example, but as you see by the graph, you must affix a minus sign to your answer for z_o. Thus z_o = - 1.2145 .

EXERCISES	ANSWERS
1) Find $P(0 \le z < 3.575)$.	1) 0.4998
2) Find $P(-2.56 \le z < 0)$	2) 0.4948
3) Find $P(-1 \le z \le 1)$	3) 2(0.3413) = 0.6826
4) Find $P(-2.345 < z < 1.96)$	4) 0.4905 + 0.4750 = 0.9655
5) If x has a normal distribution with $\mu = 54$ and $\sigma = 13.5$, find $P(x \le 68)$.	5) $P(x \le 68) = P(z \le 1.037) = .5 + .3501 = 0.8501$
6) Find the value of z, say a, such that $P(z < a) = .1789$	6) a = - .9195
7) Find the value of z, say a, such that $P(z < z_o) = .99$.	7) (tail area = .01) z_o = 2.3268

■ OVERLAYING THE NORMAL DISTRIBUTION GRAPH

In Chapter 4 of this manual, you placed the graph of the normal distribution over the graphs of the binomial, Poisson and hypergeometric probability distributions. That was accomplished using a program containing the equation of the normal density function. When you have single-variable or two-variable data entered in the STAT mode, the Sharp EL-5200 will automatically draw a normal distribution graph on the graph you have drawn using the built-in statistical graph functions.

To draw a normal distribution graph over another statistical graph already on the screen when in the data store mode, press [2ndF] [G(ND)] [AUTO] [DRAW]. This key operation causes the calculator to automatically set the value of Ymax to the value of y that results when $x = \bar{x}$ (or μ) for the normal distribution. If you wish to clear the screen and draw only the normal density function, press [2ndF] [G.CL] before the above keystroke sequence.

Normal distribution graphs can also be drawn over broken-line graphs with the keystroke sequence [2ndF] [G(ND)] [AUTO] [DRAW].

To draw a normal distribution graph in the non-store mode, press [2ndF] [G(ND)] [DRAW]. The graph will be drawn according to the contents of the range screen settings.

Return to the histogram on page 24 that was constructed for the gasoline data. Redraw this histogram and overlay a normal graph.

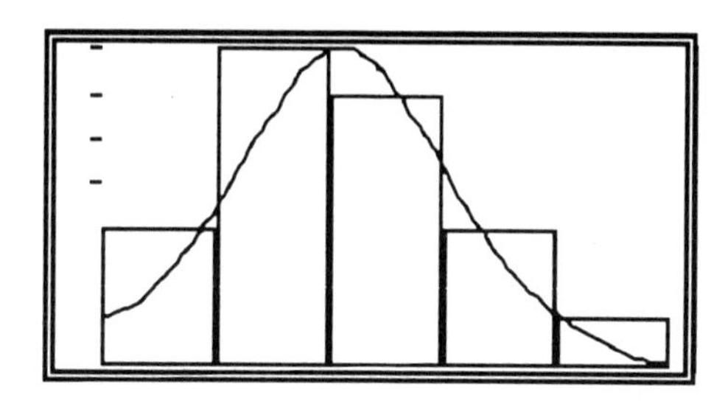

The normal distribution does not seem to approximate the distribution of the gasoline data since the normal graph does not fit closely on top of the histogram. When the normal graph does "fit" the graph of the data shown in the histogram, a normal approximation of the area between certain values on the histogram is appropriate.

EXERCISES

1) Construct histograms for n = 10, n = 50, and n = 100 tosses of 4 fair coins using the program 4 coin toss. Overlay a normal graph for each of the three cases. Do you find that the normal graph more closely fits the histogram as the value of n increases?

2) Construct histograms for n = 10, n = 50, and n = 100 tosses of 4 fair coins using the program Uniform. Overlay a normal graph for each of the three cases. Does it appear that the normal graph more closely fits the histogram as the value of n increases?

3) Compare your results of exercises 1 and 2. At what conclusion do you arrive?

CHAPTER 6

SAMPLING DISTRIBUTIONS

The process of gathering data from an experiment with chance outcomes is called sampling. The primary objective of sampling is to select a sample that is representative of the population from which it is chosen and make an inference, with a certain degree of reliability, about one or more of the unknown characteristics of that population. These population characteristics, two of which are μ and σ^2, are called population *parameters*. The values that are calculated from the sample, some of which are the sample mean $\bar{x}$, the sample median, the sample mode, and the sample variance s^2 are called sample *statistics*. The sample statistics are used to make decisions, estimates, or predictions about the population parameters.

A population parameter about which a decision is to be made is constant for any particular population. If it were possible to do a census of the population, the value of the population parameter could be exactly determined. It is usually the case, however, that a census is too costly or too time consuming. Thus, a sample is chosen from the population and a prediction of the population parameter is obtained from the information contained in that sample.

If all possible samples of size n were to be selected from a given population and if, for each of those samples, a certain statistic were computed, the value of the statistic would vary from sample to sample. The *sampling distribution* of the statistic is the probability distribution for all possible values of that statistic.

SAMPLING DISTRIBUTION OF THE SAMPLE MEAN

If it desired to make an inference about the mean μ of a population, the particular statistic used is the sample mean $\bar{x}$. If we use the notation $\mu_{\bar{x}}$ to denote the mean of all possible values of the sample mean (that is, the center of the sampling distribution of the sample mean) for a particular population and $\sigma_{\bar{x}}$ to denote the standard deviation of the sampling distribution of the sample mean, it is true that the center of the sampling distribution of the sample mean is the same as the center of the distribution of the population. That is, $\mu_{\bar{x}} = \mu$.

THE FINITE POPULATION CORRECTION FACTOR

The form of the standard deviation of the distribution of the sample mean, $\sigma_{\bar{x}}$, is dependent on how large the sample size is relative to the population size. When the population is finite of size N and the sample size is "large" relative to the size of the population, a quantity called the *finite population correction factor* should be used to compute the standard deviation of the sampling distribution of the sample mean. When the sample size is "small" relative to the population size (that is, when the population is infinite or N is very large compared to n), the value of the finite population correction factor is so close to one that it does not appreciably affect the value of the standard deviation of the sample mean. Statistics texts vary in the condition for when the finite population correction factor should be used. Check your text for the appropriate condition.

- When the finite population correction factor is not needed, $\sigma_{\bar{x}} = \dfrac{\sigma}{\sqrt{n}}$.

- When the finite population correction factor is used, $\sigma_{\bar{x}} = \frac{\sigma}{\sqrt{n}} \sqrt{\frac{N-n}{N-1}}$.

Notice that the spread of the distribution of the sample mean is less than the spread of the distribution of the population. As the sample size increases, the shape of the sampling distribution of the sample mean becomes taller and narrower.

The following program can be used to calculate the value of the standard deviation of the sampling distribution of the sample mean when the finite population correction factor should be used.

PROGRAMMING

Select the AER-II mode. You will see on the display 18: TITLE? Type in, using the letters on the right keyboard, f p c f and press [ENT] to store the program name. Input the program at the M: prompt as follows:

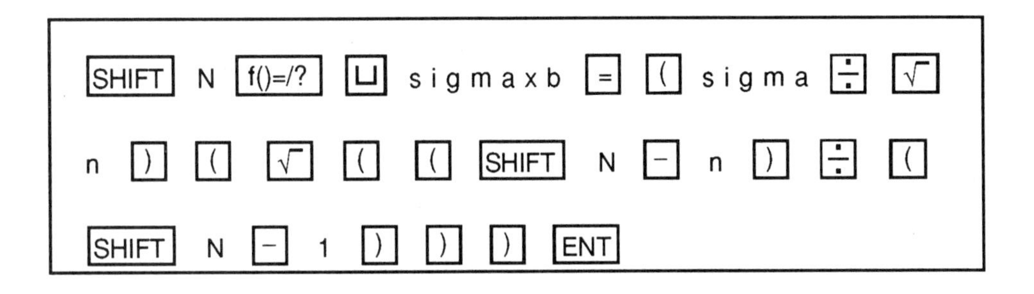

Once the program is entered, your calculator screen should look exactly like the following:

```
M: N=?␣sigmaxb=(s
igma÷√n)(√((N−n)
÷(N−1)))
```

• The letters "sigmaxb" were used to denote the standard deviation of the sampling distribution of the sample mean in the above program rather than the letters "sigmaxbar" since a variable name must be less than or equal to 7 characters.

• Now that you are putting more programs in your calculator, you may pass a particular program when searching. If so, use the second function key and the program key (2ndF TITLE PRO) to scroll backwards through the programs.

THE SHAPE OF THE SAMPLING DISTRIBUTION OF THE SAMPLE MEAN

In addition to knowing the mean and standard deviation of the sampling distribution of the sample mean, there is one more piece of information we need to know about the sampling distribution of the sample mean - its shape. Two theorems that quite often allow us to determine the form of this sampling distribution are the following:

• Whenever the sampled population has a normal probability distribution, the sampling distribution of the sample mean is a normal probability distribution.

• THE CENTRAL LIMIT THEOREM: If random samples of size n are chosen from a population, when n is large, (regardless of the distribution of the population) the sampling distribution of the sample mean is approximately normally distributed. The approximation will become more accurate as the sample size n increases.

The Central Limit Theorem is extremely important in statistical inference. Not only does it apply to the sampling distribution of the sample mean, it can be restated to apply to the sum of sample measurements which, as the sample size becomes large, would also tend to possess an approximately normal distribution.

Look back at the histograms you constructed on pages 79 and 80 for the toss of 4 fair coins. Is it clear, now that you've seen the Central Limit Theorem, why the program 4 coin toss (whose subroutine involved a sum of measurements), tended toward a mound-shaped distribution while the program Uniform (whose subroutine did not involve a sum of values), did not approach a mound-shaped distribution for increasing larger values of n?

The program Normal graphs can aid you in visualizing the relationship between the distribution of the population and the distribution of the sample mean when these are both normally distributed. Suppose random samples of size n = 25 are chosen from a normal population with mean $\mu = 35.46$ and variance $\sigma^2 = 4.32$. The sampling distribution of the sample mean will be normally distributed (why?) and will have mean $\mu_{\bar{x}} = 35.46$ and standard deviation $\sigma_{\bar{x}} = \frac{\sqrt{4.32}}{\sqrt{25}} = .4157$. Notice that the finite population correction factor is not used because the normal population is infinite. To draw the graphs, set the x range to be approximately $\mu - 3\sigma$ to $\mu + 3\sigma$ (29 to 42), Xscl = 5, Ymin = 0, Ymax = 0.6 and Yscl = 0.1. Run the program Normal graphs with the population mean and $\mu = 35.46$ and standard deviation $\sigma = \sqrt{4.32}$ and then run it again with the sampling distribution of the sample mean values $\mu_{\bar{x}} = 35.46$ and $\sigma_{\bar{x}} = .4157$. Notice the spread of the values and the center of the distribution for each of the graphs. You might also wish to explore with these graphs and see what happens to the distribution of the sample mean as the sample size increases.

EXERCISES

For each of the following problems, give

a) the shape of the sampling distribution of the sample mean
b) the mean of the sampling distribution of the sample mean
c) the standard deviation of the sampling distribution of the sample mean

1. Samples of size n = 100 are chosen from an infinite population of unknown shape with $\mu = 82.3$ and $\sigma = 8$.

2. Samples of size n = 9 are chosen from a normally distributed population with mean of 15 and variance of 18.

3. Samples of size 16 are chosen from a binomial population of size N = 96 with $\mu = 44.16$ and $\sigma^2 = 23.85$.

4. Samples of size 9 are chosen from a Poisson population with $\mu = 4$ and $\sigma^2 = 4$.

5. Samples of size 100 are chosen from a population of size N = 500. The population shape is unknown, but it is known that the population mean is 150.68 and the population variance is 64.

ANSWERS

	a)	b)	c)
1.	a) approximately normal	b) 82.3	c) 0.8
2.	a) normal	b) 15	c) 1.414
3.	a) unknown	b) 44.16	c) 1.1204
4.	a) unknown	b) 4	c) 2/3
5.	a) approximately normal	b) 150.68	c) 0.7163

CHAPTER 7

ESTIMATION AND HYPOTHESIS TESTING: SINGLE SAMPLE

INFERENCES ABOUT THE MEAN OF A POPULATION

Statistical inference is composed of two main areas: *estimation* and *tests of hypotheses.* In the process of estimation, a sample statistic satisfying certain requirements is used to predict the value of a population parameter. Hypothesis testing involves the formulation of a hypothesis about the population parameter of interest, and a decision is made the accept or reject that hypothesis on the basis of statistical evidence gathered in a sample from that population. The accuracy, or reliability of the decision, is dependent upon knowledge of the sampling distribution of the sample statistic.

Either a point estimate or an interval estimate may be given in the estimation of a population parameter. The best point estimate of the population mean μ is the sample mean $\bar{x}$. An interval estimate is a range of values within which the population parameter is estimated to lie. An interval estimate incorporates the measure of accuracy or reliability of the estimate in the formula for the interval estimate. The quantity $1 - \alpha$ is called the *confidence coefficient* for an interval estimate and the interval itself is called a $(1 - \alpha)100\%$ *confidence interval.*

The notation $z_{\alpha/2}$ is used in many of the formulas to follow. Recall that this symbol was discussed in Chapter 6. By definition, $z_{\alpha/2}$ is the value of the standard normal variable such that $P(z > z_{\alpha/2}) = \alpha/2$. Notice that the value appearing in the subscript is the right-hand tail area and thus these z-values can be obtained from the program Normal value.

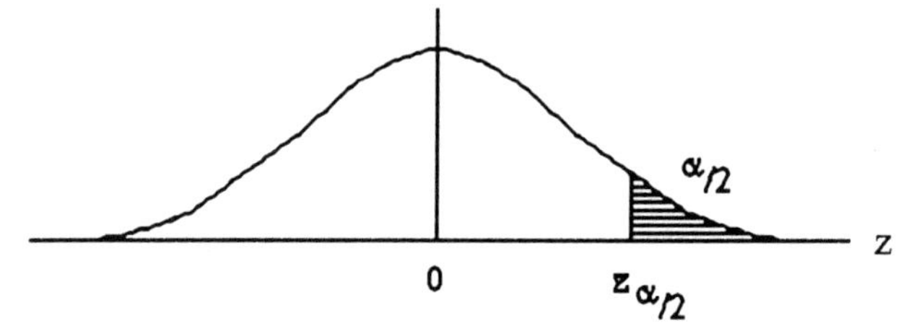

Some commonly used values of $z_{\alpha/2}$ are:

$z_{.005} = 2.575$

$z_{.01} = 2.326$

$z_{.025} = 1.960$

$z_{.05} = 1.645$

$z_{.10} = 1.282$

❒ LARGE-SAMPLE CONFIDENCE INTERVAL FOR μ

Use the following formula to give a (1-α)(100%) confidence interval for the mean μ of one population when the following conditions hold true:

- the population is normally distributed <u>or</u>
  the Central Limit Theorem applies (the standard rule of thumb is n ≥ 30)

- σ is known <u>or</u>
  s can be used to approximate σ (the approximation should only be used when n ≥ 30)

PROGRAMMING

Select the AER-II mode. You should see on the display 19: TITLE?. Type in z [␣] c i [␣] [2ndF] [SYMBOL] 6 and press [ENT] to store the program name.

Input the main routine at the M: prompt as follows:

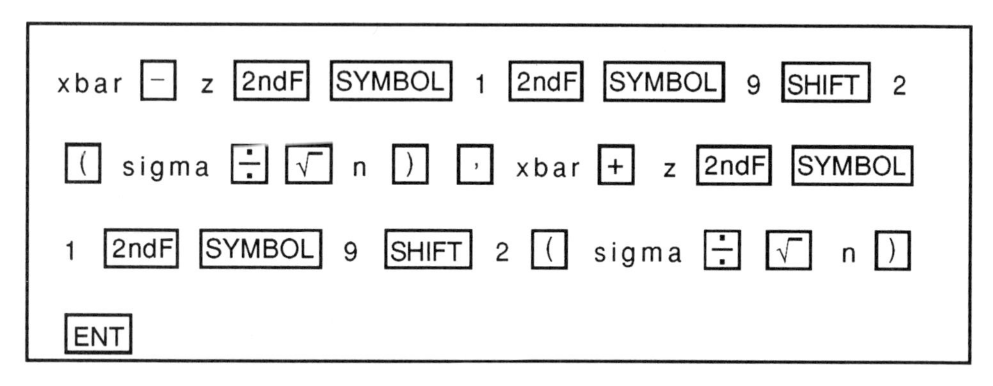

Once the program is entered, your calculator screen should appear as:

```
M: xbar − zα/2(sigm
a÷√n),xbar+zα/2(
sigma÷√n)
```

- When you go to the COMP mode to run this program, you will get ANS 1 as the first output. This is the lower endpoint of the confidence interval. Press the COMP key again and you will get ANS 2 which is the upper endpoint of the confidence interval.

EXERCISES	ANSWERS
1. A random sample of 64 items gave a sample mean of 15.67 and a standard deviation of 34. Find a 95% confidence interval for the mean of this population.	1. (7.3400, 24.000)
2. A random sample of 10 items yielded a mean of 8.0345. The population standard deviation is known to be 45.50, and the population is normally distributed. Give a 90% interval estimate of the mean of this population.	2. (-15.6344, 31.7034)

❐ LARGE-SAMPLE HYPOTHESIS TESTING FORMULA FOR μ

In the statistical process called hypothesis testing there are two hypotheses concerning a population parameter. The null hypothesis, designated by H_0, and the alternative hypothesis, designated by H_a, do not overlap and must include all possible values of the population parameter. By the way the process is formulated, hypothesis testing is designed to be supportive of the alternative hypothesis.

Because statistical inference is based on sampling procedures, there is always a chance that an error will be made in the conclusion. A type I error is made if the decision is made to reject the null hypothesis when it is actually true. The probability of a type I error is denoted by α and is called the level of significance of the test. A type II error is made if the decision is to accept the null hypothesis when it is actually false. The probability of a type II error is denoted by β. Consult your textbook for more information on hypothesis testing and the errors that are involved.

Use the following formula to give the value of the test statistic to be used in testing the hypothesis H_0: $\mu = \mu_0$ against the alternative H_a: $\mu > \mu_0$ (or $\mu < \mu_0$ or $\mu \neq \mu_0$) when the following conditions hold true:

- the population is normally distributed <u>or</u>
  the Central Limit Theorem applies (the standard rule of thumb is $n \geq 30$)
- σ is known <u>or</u>
  s can be used to approximate σ (the approximation should only be used when $n \geq 30$)

PROGRAMMING

Select the AER-II mode. You should see on the display 20: TITLE?. Type in z [␣] h t [␣] [2ndF] [SYMBOL] 6 and press [ENT] to store the program name. Input the main routine at the M: prompt as follows:

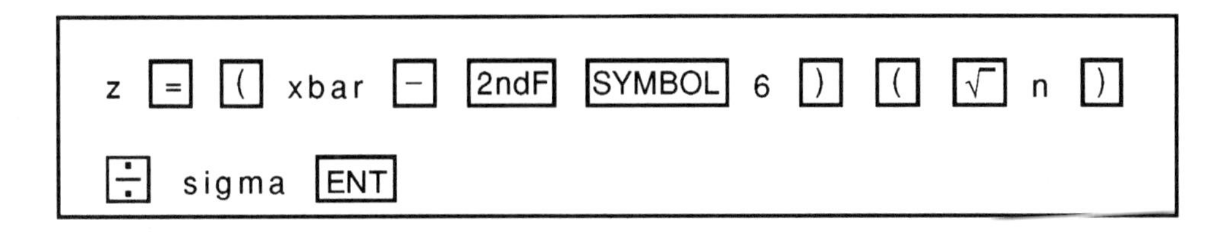

Once the program is entered, your calculator screen should look exactly like the following:

```
M: z = (xbar − μ) (√ n)
÷ sigma
```

EXERCISES

1. Test the hypothesis H_0: $\mu = 8$ against the alternative H_a: $\mu > 8$ using the following information from a random sample of size 36: the sample mean is 9.35 and the sample standard deviation is 4.56. The level of significance for the test is chosen to be 0.05.

2. A random sample of size 100 yielded a mean of 21.45 and a standard deviation of 15.82. Using P(Type I error) = .05, test H_0: $\mu = 23.5$ against the alternative hypothesis H_a: $\mu \neq 23.5$.

ANSWERS

1. $z = 1.7763$ so reject H_0
2. $z = -1.2958$ so do not reject H_0

❐ SMALL-SAMPLE CONFIDENCE INTERVAL FOR μ

Use the following formula to give a (1-α)(100%) confidence interval for the mean μ of one population when the following conditions hold true:

- the population is normally distributed
- σ is unknown.

PROGRAMMING

Select the AER-II mode. You should see on the display 21: TITLE?. Type in t [␣] c i [␣] [2ndF] [SYMBOL] 6 and press [ENT] to store the program name. Input the main routine at the M: prompt as follows:

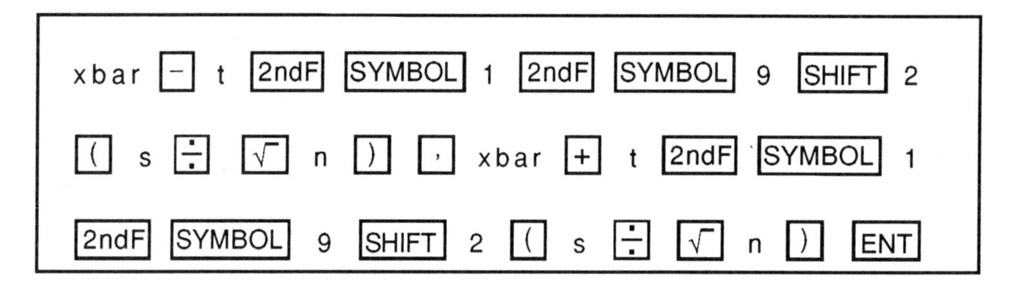

Once the program is entered, your calculator screen should appear as:

```
M: xbar – tα/2 (s ÷ √ n
) , xbar + tα/2 (s ÷ √ n
)
```

- When you go to the COMP mode to run this program, you will get ANS 1 as the first output. This is the lower endpoint of the confidence interval. Press the COMP key again and you will get ANS 2 which is the upper endpoint of the confidence interval.

EXERCISES

1. A random sample of 16 items gave a sample mean of 65 and a standard deviation of 28. Find a 95% confidence interval for the mean of this normally distributed population.
2. A random sample of 9 items yielded a sample mean of 5.0987 and a sample standard deviation of 1.0895. Give a 90% confidence interval for the mean of this normally distributed population.

ANSWERS

1. (50.0830, 79.9170)
2. (4.4232, 5.7742)

❑ SMALL-SAMPLE HYPOTHESIS TESTING FORMULA FOR μ

Use the following formula to give the value of the test statistic to be used in testing the hypothesis H_0: $\mu = \mu_0$ against the alternative H_a: $\mu > \mu_0$ (or $\mu < \mu_0$ or $\mu \neq \mu_0$) when the following conditions hold true:

- the population is normally distributed
- σ is unknown

PROGRAMMING

Select the AER-II mode. You should see on the display 22: TITLE?. Type in t [␣] h t [␣] [2ndF] [SYMBOL] 6 and press [ENT] to store the program name.

Input the main routine at the M: prompt as follows:

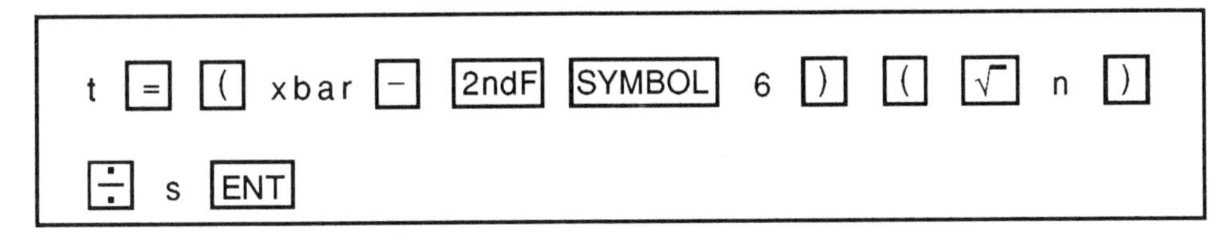

Once the program is entered, your calculator screen should look exactly like the following:

```
M: t = (xbar − μ) (√ n)
÷ s
```

EXERCISES

1. A random sample of size 24, chosen from a normal population, yielded a mean of 122.7 and a standard deviation of 12.645. Using a .02 level of significance, test H_0: $\mu = 123.5$ against the alternative H_a: $\mu < 123.5$.

2. Test H_0: $\mu = 14$ against the alternative H_a: $\mu \neq 14$ using the following results from a sample of size 22 that was chosen from a normally distributed population: sample mean = 15.6 and sample variance = 5.43. Let $\alpha = .10$.

ANSWERS

1. $t = -0.3099$ so do not reject H_0
2. $t = 1.3821$ so do not reject H_0

INFERENCES ABOUT THE BINOMIAL PROPORTION OF SUCCESS FOR ONE POPULATION

The sample proportion (also called percentage) is computed by taking the number of items in the sample with the characteristic of interest in the problem divided by the size of the random sample chosen from the population. The sample proportion is the point estimate of the population proportion. We will consider the parameter of interest to be p, the probability of success on any one trial for the binomial experiment. We consider only those cases where the normal approximation to the binomial distribution is appropriate.

LARGE-SAMPLE CONFIDENCE INTERVAL FOR p

Use the following formula to give a $(1-\alpha)(100\%)$ confidence interval for the binomial probability p of one population when the following conditions hold true:

- the population is binomially distributed
- it is appropriate to use the normal approximation to the binomial distribution; that is, the interval $\frac{x}{n} \pm 3\sqrt{\frac{(\frac{x}{n})(1-\frac{x}{n})}{n}}$ does not include 0 or 1.
- Note: The condition for when the normal approximation to the binomial distribution is "good" will vary among textbooks. Check your statistics text and modify the second condition accordingly.

PROGRAMMING

Select the AER-II mode. You should see on the display 23: TITLE?. Type in z [␣] c i [␣] p and press [ENT] to store the program name. Input the main routine as:

[(] x [÷] n [)] [−] z [2ndF] [SYMBOL] 1 [2ndF] [SYMBOL] 9 [SHIFT] 2 [√] [(] [(] x [÷] n [)] [(] 1 [−] [(] x [÷] n [)] [)] [÷] n [)] [,] [(] x [÷] n [)] [+] z [2ndF] [SYMBOL] 1 [2ndF] [SYMBOL] 9 [SHIFT] 2 [√] [(] [(] x [÷] n [)] [(] 1 [−] [(] x [÷] n [)] [)] [÷] n [)] [ENT]

Once the program is entered, your calculator screen should look exactly like the following:

```
M: (x ÷ n) − zα/2√ ( (x
÷n) (1 − (x ÷ n) ) ÷ n) ,
(x ÷ n) + zα/2 √ ((x ÷n
) (1 − (x ÷ n)) ÷ n )
```

- When you go to the COMP mode to run this program, you will get ANS 1 as the first output. This is the lower endpoint of the confidence interval. Press the COMP key again and you will get ANS 2 which is the upper endpoint of the confidence interval.

This program is quite handy when you are checking condition 2 for use of this formula. Execute this program in the COMP mode. When asked for the value of $z_{\alpha/2}$, enter the number 3. If the lower endpoint of the interval is negative, 0 is within the interval and this formula should not be used for the confidence interval. If the upper endpoint of the interval is greater than 1, then 1 is within the interval and this formula should not be used to give the confidence interval.

EXERCISES

1. A random sample of 100 shoppers at a local supermarket showed that 58 of them regularly use cents-off coupons.
   a) Is it appropriate to use the normal approximation to the binomial distribution in this problem? Why or why not?
   b) Construct a 98% confidence interval for the actual proportion of all shoppers at this supermarket who use cents-off coupons.

2. A random sample of 60 adults were surveyed and asked: Do you have a home computer? Fifteen responded that they do have a computer at home.
   a) Is it appropriate to use the normal approximation to the binomial distribution in this problem? Why or why not?
   b) Construct a 95% confidence interval for the actual proportion of all adults who have a computer at home.

3. Is the sample size $n = 500$ large enough to use the normal approximation to the binomial distribution in testing the null hypothesis H_0: $p = .985$ versus the alternative hypothesis H_a: $p > .985$?

ANSWERS

1. a) Yes, since the interval (.4319, .7281) doesn't include 0 or 1.
   b) (.4652, .6948)
2. a) Yes, since the interval (.0823, .4177) doesn't include 0 or 1.
   b) (.1404, .3596)
3. No, since 1 is contained in the interval (.9687,1.0013).

❑ LARGE-SAMPLE HYPOTHESIS TESTING FORMULA FOR p

Use the following formula to give the value of the test statistic to be used in testing the hypothesis H_0: $p = p_o$ against the alternative H_a: $p > p_o$ (or $p < p_o$ or $p \neq p_o$) when the following conditions hold true:

- the population is binomially distributed
- it is appropriate to use the normal approximation to the binomial distribution; that is, the interval $p_o \pm 3\sqrt{\frac{p_o(1 - p_o)}{n}}$ does not include 0 or 1.
- The condition for when the normal approximation to the binomial distribution is "good" will vary among textbooks. Check your statistics text and modify the second condition accordingly.

PROGRAMMING

Select the AER-II mode. You should see on the display 24: TITLE?. Type in z ⊔ h t ⊔ p and press ENT to store the program name. Input the main routine as:

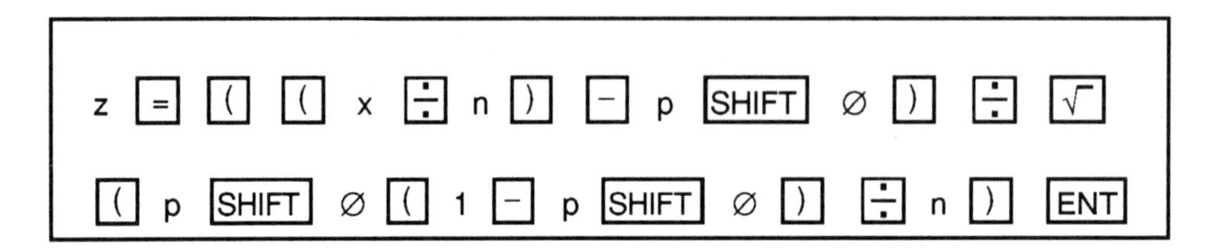

Once the program is entered, your calculator screen should look exactly like the following:

M: $z = ((x \div n) - p_o) \div \sqrt{}$
$(p_o (1 - p_o) \div n)$

- To check the condition for use of this formula, the program z ci p can be used with $z_{\alpha/2} = 3$ and $x = n(p_o)$.

EXERCISES

1. In a random sample, 15 out of 35 tennis players stated that they suffer from chronic tennis elbow.
   a) Is it appropriate to use the normal approximation to the binomial distribution?
   b) Use the 0.05 level of significance to test the hypothesis that the proportion of all tennis players who suffer from tennis elbow is less than 50 percent.

2. To check an airline's claim that more than 80% of lost luggage is returned to the rightful owner within one year, a random sample of 100 owners of lost luggage were surveyed. After a period of one year it was found that 91 owners had received their luggage sometime during that year.
   a) Is it appropriate to use the normal approximation to the binomial distribution?
   b) Use the 0.01 level of significance to test the hypothesis that the proportion of all owners of lost luggage on this airline received their luggage within one year.

3. Is the sample size n = 100 large enough to use the normal approximation to the binomial distribution in testing the hypothesis H_o: $p = .05$ versus H_a: $p \neq .05$?

ANSWERS

1. a) Yes, since the interval (.2465, .7535) does not contain 0 or 1.
   b) $z = -0.8452$ so do not reject H_0

2. a) Yes, since the interval (.6800, .9200) does not contain 0 or 1.
   b) $z = 2.7500$ so reject H_0

3. No, since 0 is contained in the interval (-.0154, .1154).

DETERMINING THE SAMPLE SIZE

One question that arises in statistical inference is that of determining how large a random sample is needed to achieve a desired degree of accuracy. An experimenter will decide on a certain reliability of the inference, a bound that should not be exceeded, and some estimate of the population standard deviation when that quantity is not known. The appropriate sample size can then be determined.

DETERMINING THE SAMPLE SIZE NECESSARY TO MAKE INFERENCES ABOUT A POPULATION MEAN

The formula used to determine the sample size necessary to make an inference about a population mean is $n = \frac{z^2_{\frac{\alpha}{2}} \sigma^2}{B^2}$. The quantity $z_{\alpha/2}$ is the value of the standard normal distribution determined from the inference confidence of $(1 - \alpha)100\%$, B is the bound that should not be exceeded, and σ is the population standard deviation. The value of n must be a whole number, so the numerical value determined by the above formula must be *rounded up* to the next highest whole number.

If the population standard deviation is unknown, an estimate for it may be obtained from the results of prior sampling or from the range of the data that one expects to get in the sample. The Empirical Rule states that for mound-shaped data, approximately 95% of the data should be within two standard deviations of the mean and approximately all of the data should fall within three standard deviations of the mean. If the shape of the data is unknown, Chebyshev's theorem states that at least 75% of the data should fall within two standard deviations of the mean and at least 88.89% of the data should fall within three standard deviations of the mean. For most data sets, two standard deviations on either side of the mean (a range of 4σ) will enclose practically all of the data. If you wish a more conservative estimate, you can use use three standard deviations on either side of the mean (a range of 6σ).

PROGRAMMING

Select the AER-II mode and you should see the display 25: TITLE?. Type in, using the letters on the right keyboard, [SHIFT] S a m p l e [␣] s i z e [␣] [2ndF] [SYMBOL] 6 and press [ENT] to store the program name.

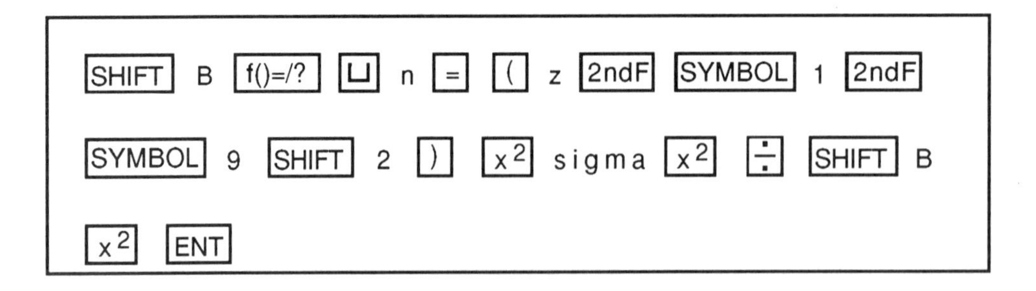

Once the program is entered, your calculator screen should look exactly like the following:

```
M: B = ?␣ n = (zα/2)² s
igma² ÷ B²
```

- If you are going to input many programs into your calculator, it is a good idea to capitalize the first letter of a program name to more easily search for a particular program. For example, to execute this program, after placing the calculator in the COMP mode, type S and press [PRO]. The display will show, in ascending number, the names of all programs beginning with S. When the desired program is found, press [COMP] to begin execution.

- To change the first letter of any of your previously entered programs, have the program name on the screen in the mode in which the program was created, press [▷] to move the cursor over the letter you wish changed and press [SHIFT] followed by the letter you wish capitalized.

EXERCISES	ANSWERS
1) Determine the sample size necessary to obtain an interval estimate of μ to within 0.50 with confidence of 95% if it is known that the population standard deviation is approximately 28.	1) 12,048
2) An experimenter wishes estimate the mean of a population. He has no prior estimate of the population variance, but he expects the data values to fall between 45 and 65. If outliers are not expected and he wishes to estimate μ to within 1 withprobability of 0.95, how large a sample should he take?	2) 97

DETERMINING THE SAMPLE SIZE NECESSARY TO MAKE INFERENCES ABOUT A POPULATION PROPORTION

The formula used to determine the sample size necessary to make an inference about a population probability (percentage) is $n = \dfrac{z^2_{\frac{\alpha}{2}}\,(p)\,(1-p)}{B^2}$. The quantity $z_{\alpha/2}$ is the value of the standard normal distribution determined from the inference confidence of $(1 - \alpha)100\%$, B is the bound that should not be exceeded, and p is an estimate of the binomial population probability of success. The value of n must be a whole number, so the numerical value determined by the above formula must be *rounded up* to the next highest whole number.

PROGRAMMING

Select the AER-II mode and you should see the display 26: TITLE?. Type in, using the letters on the right keyboard, SHIFT S a m p l e ⊔ s i z e ⊔ p and press ENT to store the program name. Input the following program as:

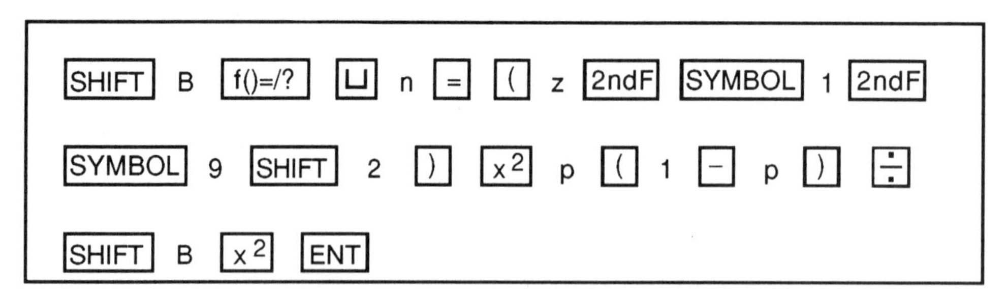

Once the program is entered, your calculator screen should look exactly like the following:

M: B = ? ⊔ n = (zα/2) 2 p
(1 − p) ÷ B2

AN APPLICATION TO CALCULUS

If no estimate of p is known, what value should be used in the formula to determine the sample size necessary for the experiment? The answer is the value that gives the maximum sample size for a fixed $z_{\alpha/2}$ and B. Thus, we wish to find the value of p which maximizes the function $p(1-p) = p - p^2$.

Because the EL-5200 uses x as the independent variable for graphing, let us rewrite the problem as: Find the value of x which maximizes $y = x - x^2$. Place the calculator in the COMP mode, press [RANGE] [2ndF] [CL] [RANGE] [RANGE] (to set the default range parameters) and enter [GRAPH] X [−] X [x^2] [DRAW]. Recall that while in the graphics screen, to use the tracing function you press and hold [▷] as the blinking dot moves along the curve. You can find the corresponding value of y at a particular value of x by pressing [2ndF] [X⇔Y]. Use the tracing function to find the value of x at which the maximum of $y = x - x^2$ occurs. You should find the maximum of $y = 0.25$ occurring at $x = 0.5$. Thus, if no estimate of p, the binomial probability of success, is available, you should use $p = 0.5$.

EXERCISES	ANSWERS
1) Before a bill to institute a state lottery with the proceeds going to education comes before a state senate, a congressman would like to know how his constitutents feel about the issue. Approximately how many voters should be surveyed to estimate the true proportion favoring this bill to within .05 with probability equal to .99?	1) 664
2) Dr. Martin believes that approximately 20% of his patients have serious common colds during the month of January. If he wishes to estimate the true percentage to within 10%, how large a random sample of his patients should he take? Assume that he wishes to be 95% confident of his answer.	2) 62

CHAPTER 8

ESTIMATION AND HYPOTHESIS TESTING: TWO SAMPLES

INFERENCES ABOUT THE MEANS OF TWO POPULATIONS

Many instances arise when an interval estimate of the difference in the means of two populations is necessary. These interval estimates and decisions made as a result of tests of hypotheses concerning the difference of two populations means are based on the information contained in two random samples, one chosen from each population. Samples are considered *independent* when they are completely separate and unrelated and the results of one sample in no way affect the results of the other.

LARGE-SAMPLE CONFIDENCE INTERVAL FOR $\mu_1 - \mu_2$

Use the following formula to give a $(1-\alpha)(100\%)$ confidence interval for $\mu_1 - \mu_2$, the difference between the means of population 1 and population 2, when the following conditions hold true:

- both populations are normally distributed <u>or</u>
  the Central Limit Theorem applies (the general rule is $n_1 \geq 30$, $n_2 \geq 30$)
- σ_1 and σ_2 are known <u>or</u>
  s_1 can be used to approximate σ_1 ($n_1 \geq 30$) and s_2 can be used to approximate σ_2 ($n_2 \geq 30$)
- the two samples are randomly selected in an independent manner from the two populations being sampled.

Conditions for use of the statistical formulas that are presented in this manual are the ones commonly given in most statistics textbooks. Verify that these are the ones appropriate for your class with your instructor.

PROGRAMMING

Select the AER-II mode. You should see on the display 27: TITLE?. Type in z [␣] c i [␣] [2ndF] [SYMBOL] 6 [SHIFT] 1 [−] [2ndF] [SYMBOL] 6 [SHIFT] 2 and press [ENT] to store the program name. Input the main routine at the M: prompt as follows:

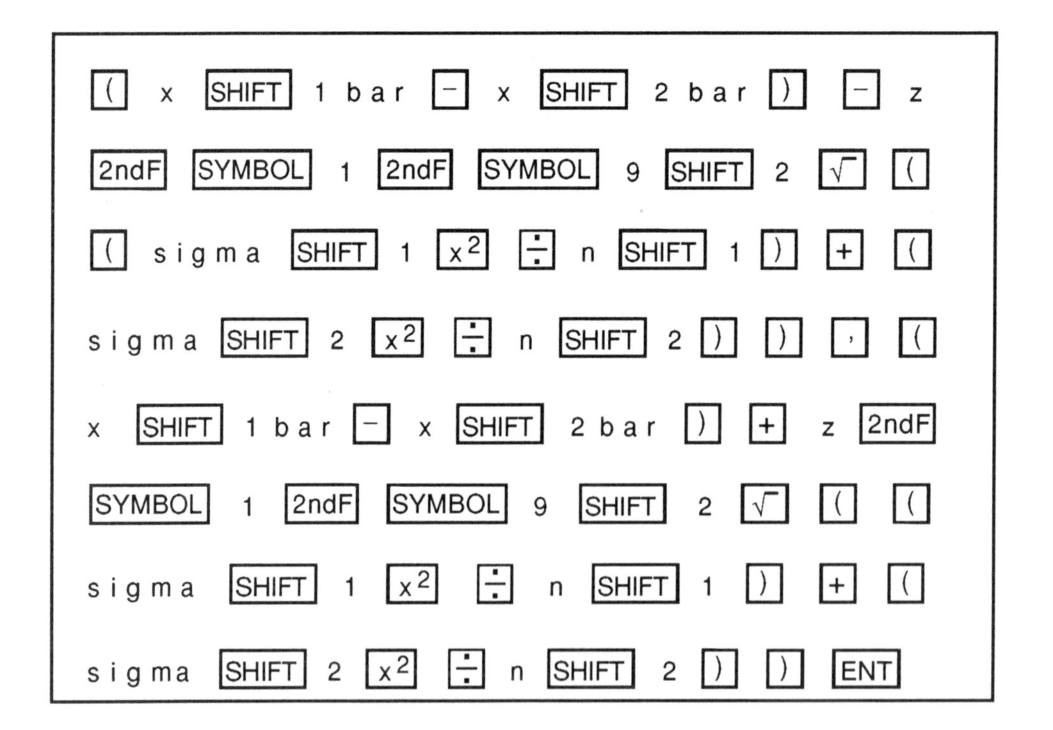

Once the program is entered, your calculator screen should look exactly like:

M: ($x_1$bar − $x_2$bar) −
z$\alpha/2$ √ ((sigma$_1{}^2$ ÷ n
$_1$) + (sigma$_2{}^2$ ÷ n_2))
, ($x_1$bar − $x_2$bar) + z
$\alpha/2$ √ ((sigma$_1{}^2$ ÷ n_1
) + (sigma$_2{}^2$ ÷ n_2))

- When you go to the COMP mode to run this program, you will get **ANS 1** as the first output. This is the lower endpoint of the confidence interval. Press the COMP key again and you will get **ANS 2** which is the upper endpoint of the confidence interval.

EXERCISES

1. A comparison of the yield of two different varieties of corn, Country Gentlemen and Silver Queen, was obtained by planting and growing 50 acres of each type under similar growth conditions. The Country Gentlemen variety yielded 80 bushels per acre with a standard deviation of 5 bushels per acre. The Silver Queen variety yielded 75 bushels per acre with a standard deviation of 6 bushels per acre. Estimate the mean difference in yield per acre between the two varieties with a 90% confidence interval and interpret your answer.

2. Consider the following data obtained from two independently chosen samples from two different populations:

Sample 1	mean = 7.5	variance = 1.5	sample size = 60
Sample 2	mean = 10.6	variance = 2.9	sample size = 35

   a) Construct a 98% confidence interval for the difference in the means of the two populations.
   b) Would a 90% confidence interval for the difference in the two population means, constructed using the same sample sizes, be narrower or wider than the interval constructed in a)?
   c) Suppose the sample sizes for both samples were increased to 100. Would a 90% confidence interval for the difference in the two population means be narrower or wider than the one constructed in b)?

ANSWERS

1) The mean for the Country Gentlemen acres minus the mean for the Silver Queen acres is estimated to be in the interval (3.1380, 6.8170).
   One interpretation that can be given for this interval estimate is that the person doing the comparison can be 90% certain that the mean yield per acre of land planted with Country Gentlemen corn is between 3.1380 and 6.8170 bushels higher than the mean yield per acre of land planted with Silver Queen corn.
   Another interpretation, this one on the confidence, is that out of many such intervals constructed in a similar manner, the above-mentioned difference will be contained in 90% of the intervals; this difference will not be in 10% of the intervals.
2) a) $\mu_1 - \mu_2$ is estimated to be in the interval (-3.8646, -2.3354).
   b) We would expect the interval to be narrower since the confidence is decreased. Note that the resulting interval is (-3.6402, -2.5598).
   c) We would expect the interval to be narrower since more information is obtained about the true population difference with the larger sample sizes. The resulting interval is (-3.4451, -2,7549).

❑ LARGE-SAMPLE HYPOTHESIS TESTING FORMULA FOR $\mu_1 - \mu_2$

Use the following formula to give the value of the test statistic to be used in testing the hypothesis H_o: $\mu_1 - \mu_2 = D$ against the alternative H_a: $\mu_1 - \mu_2 > D$ (or $\mu_1 - \mu_2 < D$ or $\mu_1 - \mu_2 \neq D$) when the following conditions hold true:

- both populations are normally distributed <u>or</u>
  the Central Limit Theorem applies (the general rule is $n_1 \geq 30$, $n_2 \geq 30$)

- σ_1 and σ_2 are known <u>or</u>
  s_1 can be used to approximate σ_1 ($n_1 \geq 30$) and s_2 can be used to approximate σ_2 ($n_2 \geq 30$)

- the two samples are randomly selected in an independent manner from the two populations being sampled.

PROGRAMMING

Select the AER-II mode. You should see on the display 28: TITLE?. Type in z [␣] h t [␣] [2ndF] [SYMBOL] 6 [SHIFT] 1 [−] [2ndF] [SYMBOL] 6 [SHIFT] 2 and press [ENT] to store the program name. Input the main routine at the M: prompt as follows:

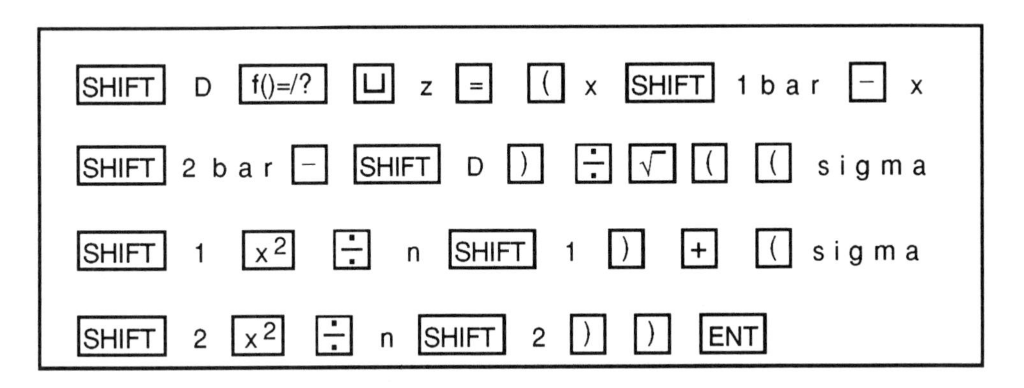

Once the program is entered, your calculator screen should look exactly like:

```
M: D=?␣ z=(x1bar−x
2bar −D)÷√((sigma
1^2 ÷ n1) + (sigma2^2 ÷
n2))
```

- The "Shift D f()=/? " keystroke sequence defines the capital letter D as a variable and prompts for the value of D when the program is executed.

EXERCISES

1. A student wishes to determine if he actually saves money by buying the cost-cutter VCR tape rather than the leading brand VCR tape. He randomly and independently selects two samples and determines the number of hours before snow or streaks occur in the playback picture for each of the brands of tape. For 50 tapes of the cost-cutter brand a mean of 65.3 hours and a standard deviation of 8.5 hours are found. The mean and standard deviation for 50 tapes of the leading brand tape are determined to be 74.6 hours and 10.4 hours, respectively. Test the hypothesis that the cost cutter tapes do not last as long in terms of mean life before snow or streaks occur in the playback picture. Use $\alpha = 0.10$.

2. Test the hypothesis H_0: $\mu_1 - \mu_2 = 8$ against the alternative H_a: $\mu_1 - \mu_2 > 8$ using the following information:

Sample 1	mean = 17.3	variance = 12.74	sample size = 50
Sample 2	mean = 8.6	variance = 15.29	sample size = 45

   Use a level of significance of 0.05.

ANSWERS

1. Letting "1" refer to the cost cutter tape and "2" reference the leading brand VCR tape, we are testing H_0: $\mu_1 - \mu_2 = 0$ against the alternative H_a: $\mu_1 - \mu_2 < 0$ or equivalently, H_a: $\mu_1 < \mu_2$. The null hypothesis will be rejected whenever $z < -1.282$. Since the test statistic, z, equals -4.896, it falls in the rejection region. Thus, we reject H_0.

2. Since the rejection region is $z > 1.645$ and $z = 0.9078$, the conclusion is do not reject H_0. There is insufficient evidence to believe that $\mu_1 - \mu_2 > 8$ at the .05 level of significance.

❑ SMALL-SAMPLE CONFIDENCE INTERVAL FOR $\mu_1 - \mu_2$

Small-sample inferences about the difference between population means can use either an independent-sample design (in which case the results of one sample do not affect the results of the other sample) or a matched-sample design. When observations in one sample are paired or matched with observations in the other sample, the samples are not independent and are called *paired* or *matched*.

The independent-sample design formula employs a pooled estimator of the population variance and is therefore different from the formula used to find confidence intervals or perform hypothesis tests in the large-sample case. Let's first consider the case where the samples are independently and randomly chosen from the populations of interest. We will then consider the paired samples situation which employs difference data.

Use the following formula to give a (1-α)(100%) confidence interval for $\mu_1 - \mu_2$, the difference between the means of population 1 and population 2, when the following conditions hold true:

- both populations are normally distributed
- σ_1 and σ_2 are unknown, but $\sigma_1^2 = \sigma_2^2$
- the two samples are randomly selected in an independent manner from the two populations being sampled.

PROGRAMMING

Select the AER-II mode. You should see on the display 29: TITLE?. Type in t [␣] c i [␣] [2ndF] [SYMBOL] 6 [SHIFT] 1 [−] [2ndF] [SYMBOL] 6 [SHIFT] 2 [␣] [␣] [␣] i n d p t [␣] s a m p l e s and press [ENT] to store the program name. Input the main routine at the M: prompt as follows:

[2ndF] 1 [,] [(] x [SHIFT] 1 b a r [−] x [SHIFT] 2 b a r [)] [−] t [2ndF] [SYMBOL] 1 [2ndF] [SYMBOL] 9 [SHIFT] 2 [√] spoolsq [(] [(] 1 [÷] n [SHIFT] 1 [)] [+] [(] 1 [÷] n [SHIFT] 2 [)] [)] [,] [(] x [SHIFT] 1 b a r [−] x [SHIFT] 2 b a r [)] [+] t [2ndF] [SYMBOL] 1 [2ndF] [SYMBOL] 9 [SHIFT] 2 [√] spoolsq [(] [(] 1 [÷] n [SHIFT] 1 [)] [+] [(] 1 [÷] n [SHIFT] 2 [)] [)] [2ndF] [SUB] *(at the* [1]: *prompt, type the following)* spoolsq [=] [(] [(] n [SHIFT] 1 [−] 1 [)] s [SHIFT] 1 [x^2] [+] [(] n [SHIFT] 2 [−] 1 [)] s [SHIFT] 2 [x^2] [)] [÷] [(] n [SHIFT] 1 [+] n [SHIFT] 2 [−] 2 [)] [ENT]

Once the program is entered, your calculator screen should look exactly like:

M: [1] , ($x_1$bar− $x_2$bar
) − tα/2 √ spoolsq ((
1 ÷ n_1) + (1÷ n_2)) , (x
$_1$bar − x_2 bar) + tα/2
√ spoolsq ((1÷ n_1) +
(1 ÷ n_2))

Press [2ndF] [▽] to see the contents of subroutine 1 which should appear as:

```
[1]: spoolsq = ((n1-1
) s1² + (n2-1) s2²) ÷
(n1 + n2 -2)
```

- This program uses a subroutine to calculate the value of the pooled estimate (spoolsq = s_p^2) of the common population variance. Press [COMP] and you will display ANS2 which is the lower endpoint of the confidence interval. Press the [COMP] key again and you will display ANS3 which is the upper endpoint of the confidence interval. (These are labeled ANS2 and ANS3 because ANS1 is the value of s_p^2 which is given as "spoolsq" when the program is being executed. If you do not wish to have the value of spoolsq displayed, omit the comma after the " [2ndF] 1 " at the beginning of the program.)

The t distribution is appropriate to use in any statistical inference concerning population means when the above conditions hold true. However, due to the fact that for samples of size thirty or more, t and z are almost identical, we usually use the z formulas with the population standard deviations estimated by the sample standard deviations to develop confidence intervals and test hypotheses. By doing this, we are not restricted by the required assumption of normally distributed populations for the t distribution. In the small-sample case, however, we must use t.

❐ SMALL-SAMPLE HYPOTHESIS TEST FOR $\mu_1 - \mu_2$

Use the following formula to give the value of the test statistic to be used in testing the hypothesis H_0: $\mu_1 - \mu_2 = D$ against the alternative H_a: $\mu_1 - \mu_2 > D$ (or $\mu_1 - \mu_2 < D$ or $\mu_1 - \mu_2 \neq D$) when the following conditions hold true:

- both populations are normally distributed
- σ_1 and σ_2 are unknown, but $\sigma_1^2 = \sigma_2^2$
- the two samples are randomly selected in an independent manner from the two populations being sampled.

PROGRAMMING

Select the AER-II mode. You should see on the display 30: TITLE?. Type in t [⊔] h t [⊔] [2ndF] [SYMBOL] 6 [SHIFT] 1 [−] [2ndF] [SYMBOL] 6 [SHIFT] 2 [⊔] [⊔] [⊔] i n d p t [⊔] s a m p l e s and press [ENT] to store the program name. Input the main routine at the M: prompt as follows:

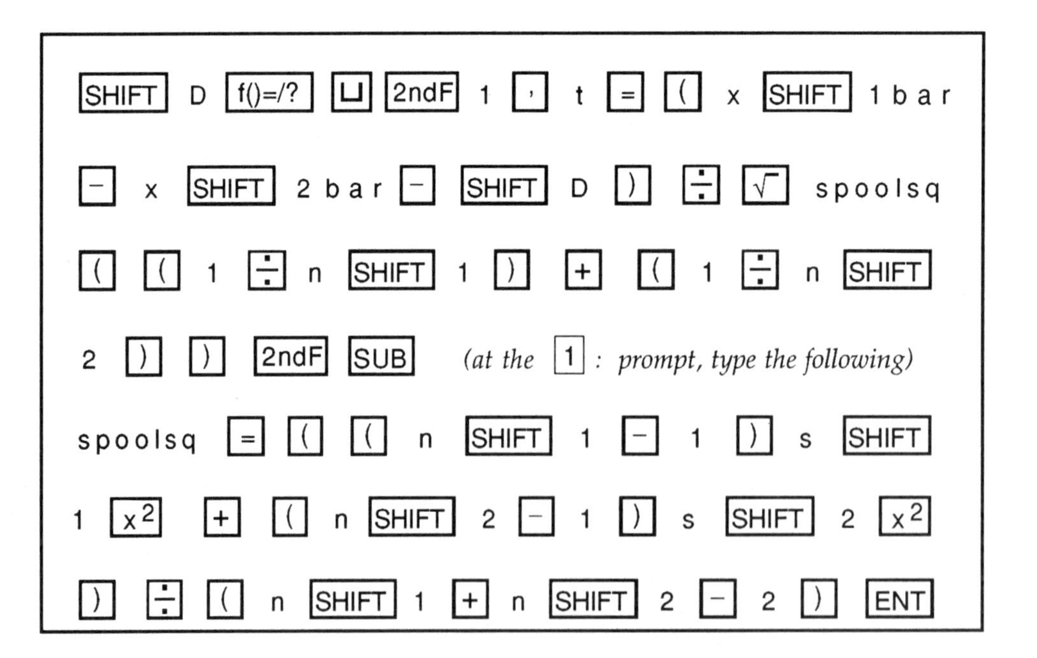

Once the program is entered, your calculator screen should look exactly like:

```
M: D=?␣[1], t = (x1bar
- x2bar - D) ÷ √ spool
sq ((1 ÷ n1) + (1 ÷ n2)
)
```

Press [2ndF] [▽] to see the contents of subroutine 1 which should appear as:

```
[1]: spoolsq = ((n1 -1
) s1² + (n2 - 1) s2² ) ÷
(n1 + n2 - 2)
```

You may have noticed that for use of this formula the condition of σ_1 and σ_2 being unknown but $\sigma_1^2 = \sigma_2^2$, seems quite difficult to check. There is a statistical test of hypothesis called the F test which is used to determine whether the condition of equal population variances is appropriate. Check your statistics text for the details of this hypothesis test.

EXERCISES

1. A dentist wishes to estimate, with a 99% confidence interval, the difference in the mean number of cavities in children who have a fluoride treatment at the time of their cleaning and the mean number of cavities in children who do not have the treatment. Fifteen children who had fluoride treatments over a period of one year were compared to 8 children who did not. Of the fifteen receiving the treatment, the mean number of cavities was 1.56 with a standard deviation of 0.89. For those eight not receiving the fluoride, the mean number of cavities was 2.41 with a standard deviation of 0.95.
   a) Give the interval estimate.
   b) Are there any assumptions that must be made in order for your estimate to be valid?

c) Test the hypothesis that the children who have the fluoride treatment have fewer cavities that the children who do not have the treatment. Use a level of significance of 0.05.

d) Would your decision to c) have changed if $\alpha = 0.01$ instead of 0.05?

2. Independent random samples of the gpa's (grade point averages) of sixteen sophomores and thirteen juniors attending a large state university gave the statistics below for cumulative grade point averages:

Class	Sample Mean	Sample Variance	Sample Size
(1) Sophomores	2.54	0.18	16
(2) Juniors	2.75	0.15	13

The populations from which the samples were drawn are known to be approximately normally distributed.

a) Construct a 95% interval estimate of the true difference in the gpa's of all sophomores and juniors at this university.

b) Can you conclude from these data that there is a difference in the mean cumulative grade point averages for all sophomores and juniors at the university? Use $\alpha = 0.05$.

ANSWERS

1. a) You should have $s_p^2 = 0.8289$. The confidence interval is (-1.9784, 0.2784).

   b) Yes, we must assume that both populations are normally distributed with equal variances and that the samples are randomly and independently chosen from those populations.

   c) Letting "1" refer to the fluoride treatment children and "2" refer to the children who do not receive the treatment, we are testing H_0: $\mu_1 - \mu_2 = 0$ against the alternative H_a: $\mu_1 - \mu_2 < 0$ ($\mu_1 < \mu_2$). The null hypothesis is rejected whenever $t < -1.721$. Since the test statistic, t, equals -2.1325, it falls in the rejection region. Thus, we reject H_0 and conclude that the children with the fluoride treatment have significantly fewer cavities than the children without the fluoride treatment.

   d) The null hypothesis would be rejected whenever $t < -2.518$. Yes, using α equal to 0.01, you should come to the conclusion that there is *not* sufficient evidence to believe the alternative hypothesis.

2. a) You should have $s_p^2 = 0.1667$. The interval estimate is (-0.5228, 0.1028).

   b) We are testing H_0: $\mu_1 - \mu_2 = 0$ against the alternative H_a: $\mu_1 - \mu_2 \neq 0$ ($\mu_1 \neq \mu_2$). The null hypothesis is rejected whenever $t < -2.052$ or $t > 2.052$. Since $t = -1.3776$, the test statistic does not fall in the rejection region. Thus, we reject do *not* reject H_0.

❐ PAIRED SAMPLES

Problems involving paired or matched samples require the entry of the differences in the original data, not the summary statistics. To use the programs for paired samples, you must perform the following steps for entering the data from the two samples *before* executing either of the programs given for paired samples.

- Place the calculator in the STAT mode and choose data store.

- Press TITLE DATA and with the blinking cursor on the S, press 2ndF CL and ENT to clear memory array S. Press 2ndF T-G-D to return to the text screen.

- Enter the *data differences* by using the numeric keys, the subtraction key − and the (white) data key DATA . (For example, to enter the paired values 3 and 1, key in 3, press − , key in 1 and press DATA .) You will see the difference of the values appearing on the left of the text screen. Recall that the values on the right of the text screen give the number of data values that have been entered.

- After all data differences are entered, place the calculator in the COMP mode to execute the program.

Recall that the number of data values that are entered, the sum of the entered values and the sum of the squares of the entered data (see page 32 of this manual) are stored in memory array Z. (If you wish to see which values are stored in memory location Z, you must place the calculator in the COMP mode, press TITLE DATA and use

▽ to position the blinking cursor over the "Z". Press = and the stored values will appear on the right of the screen.) This data can be used in a program by accessing it with the keystrokes "Z [#] " where # is the numerical value specifying the location of the desired information.

The confidence interval formula and the hypothesis testing formula for paired data are exactly the same equations that are used in giving inferences for the mean of a single population. However, instead of data being obtained from a single sample, the difference data from two paired samples is used. To access the data stored in memory location Z the formulas must be rewritten in terms of x where x is $x_1 - x_2$, the difference obtained from the data in the two paired samples.

Recall that $\bar{x} = \frac{\sum x}{n}$ and $s^2 = \frac{\sum x^2 - \frac{(\sum x)^2}{n}}{n-1}$. Algebraically, it can be shown that $\frac{s}{\sqrt{n}}$ equals $\frac{\sqrt{n \sum x^2 - (\sum x)^2}}{n\sqrt{n-1}}$ and that the hypothesis testing formula $t = \frac{\bar{x} - D}{\frac{s}{\sqrt{n}}}$ can be written as $t = \frac{\sqrt{n-1}\,(\sum x - nD)}{\sqrt{n \sum x^2 - (\sum x)^2}}$. These algebraically equivalent formulas are the ones that are given in the programming formulas for the confidence interval and hypothesis test for paired samples.

PAIRED SAMPLES CONFIDENCE INTERVAL FOR $\mu_1 - \mu_2$

Use the following formula to give a $(1\text{-}\alpha)(100\%)$ confidence interval for $\mu_1 - \mu_2 = \mu_D$, the difference between the means of population 1 and population 2, when the following conditions hold true:

- the observations in the samples are paired or matched
- the population of differences is approximately normally distributed.

PROGRAMMING

Select the AER-II mode. You should see on the display 31: TITLE?. Type in t [␣] c i [␣] [2ndF] [SYMBOL] 6 [SHIFT] 1 [−] [2ndF] [SYMBOL] 6 [SHIFT] 2 [␣] [␣] [␣] p a i r e d [␣] s a m p l e s and press [ENT] to store the program name. Input the main routine at the M: prompt as follows:

- Note: In the following program, " l c l " and " u c l " are obtained using the lower case alphabet *letters* on the right keyboard.

- Two subroutines are used here to avoid having to type in the same lengthy expression twice. Instead of displaying ANS1 and ANS2 as the respective lower and upper endpoints of the interval, you will see "l c l " for the lower limit and "u c l " for the upper limit of the confidence interval.

2ndF 1 2ndF 2 l c l = a − b , u c l = a + b 2ndF SUB *(at the* 1: *prompt, type the following)* a = (SHIFT Z 2ndF [2 2ndF] ÷ SHIFT Z 2ndF [1 2ndF]) 2ndF SUB *(at the* 2: *prompt, type the following)* b = t 2ndF SYMBOL 1 2ndF SYMBOL 9 SHIFT 2 √ (SHIFT Z 2ndF [1 2ndF] SHIFT Z 2ndF [3 2ndF] − SHIFT Z 2ndF [2 2ndF] x^2) ÷ (SHIFT Z 2ndF [1 2ndF] √ (SHIFT Z 2ndF [1 2ndF] − 1)) ENT

Once the program is entered, your calculator screen should appear as:

M: 1 2 lcl=a−b,ucl= a+b

Press 2ndF ▽ to see the contents of subroutine 1 which should appear as:

1: a = (Z[2] ÷ Z[1])

Press 2ndF ▽ to see the contents of subroutine 2 which should appear on your screen exactly like this:

```
2 : b = tα/2 √ (Z [1] Z [
3] - Z[2]² ) ÷ (Z[1] √
( Z[1] -1) )
```

❐ PAIRED SAMPLES HYPOTHESIS TEST FOR $\mu_1 - \mu_2$

Use the following formula to give the value of the test statistic to be used in testing the hypothesis H_0: $\mu_1 - \mu_2 = D$ against the alternative H_a: $\mu_1 - \mu_2 > D$ (or $\mu_1 - \mu_2 < D$ or $\mu_1 - \mu_2 \neq D$) when the following conditions hold true:

- the observations in the samples are paired or matched
- the population of differences is approximately normally distributed.

The programs for the paired sample confidence interval and the paired sample hypothesis test may also be used to obtain confidence intervals and hypothesis test results for inferences about a single population whenever raw data instead of summary statistics are given. Simply enter the data in the data store mode of the STAT mode with the M+ key, change to the COMP mode and run the programs for the paired sample confidence interval and the paired sample hypothesis test using the original one sample data rather than the difference data.

PROGRAMMING

Select the AER-II mode. You should see on the display 32: TITLE?. Type in t [␣] h t [␣] [2ndF] [SYMBOL] 6 [SHIFT] 1 [−] [2ndF] [SYMBOL] 6 [SHIFT] 2 [␣] [␣] [␣] p a i r e d [␣] s a m p l e s and press [ENT] to store the program name. Input the main routine at the M: prompt as follows:

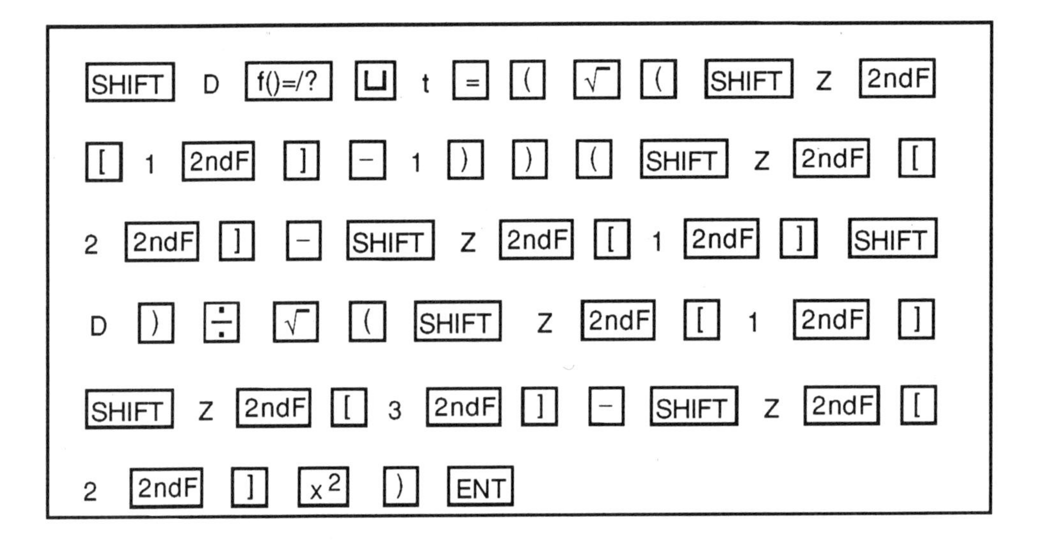

Once the program is entered, your calculator screen should appear as:

```
M: D=?␣t = ( √ ( Z[1]−
1 ) ) (Z[2] − Z[1] D ) ÷
√ ( Z[1 ] Z[ 3 ] − Z[ 2 ]²
)
```

- Remember that the data for paired difference inferences must be entered in the STAT mode before you execute the program in the COMP mode.

FOOD FOR THOUGHT

When testing the hypothesis H_0: $\mu_1 - \mu_2 = 0$, we can visually check the conclusion at which we arrive through the use of the G(SD) function in the calculator. This function will plot a scatter diagram of the data pairs. (Refer to page 170 of this manual for the procedure to enter two-variable data.) Let's call the data from sample 1 the x values and the data from sample 2 the y values. After entering the (x,y) pairs, press 2ndF G(SD) AUTO DRAW. Next overlay the line $y = x$ on the data with the keystrokes GRAPH X DRAW.

If the points appear close to the line, this would lead us to believe that $x = y$ ($\mu_1 = \mu_2$). If most of the data points are above the line, we would suspect that $x < y$ ($\mu_1 < \mu_2$). If most of the data points are below the line, we would be lead to believe that $x > y$ ($\mu_1 > \mu_2$).

EXERCISES

1. The pulse rates of a random sample of six patients before and after being given a certain tranquilizer are as follows:

Person	Pulse Rate Before Tranquilizer	Pulse Rate After Tranquilizer
1	81	77
2	80	79
3	78	75
4	79	80
5	84	79
6	80	74

a) Give a 98% confidence interval for the true difference in the patient pulse rate before and after being given this tranquilizer.

b) Using $\alpha = 0.01$ and the paired data above, can it be concluded that the use of the tranquilizer reduces the pulse rate of all the patients given this tranquilizer?

2. Suppose that the experimenter decides to increase the size of the sample to see what effect this would have on the results, and suppose she decides to extend the study to include 5 more patients. The results are:

Person	Pulse Rate Before Tranquilizer	Pulse Rate After Tranquilizer
7	83	79
8	82	79
9	75	76
10	88	83
11	81	80

a) Give a 98% confidence interval, using the combined data, for the true difference in the patient pulse rate before and after being given this tranquilizer .
b) Using $\alpha = 0.01$ and the combined paired data, can it be concluded that the use of the tranquilizer reduces the pulse rate of all patients?
c) Did the increased sample size change the result of the hypothesis test?
d) Draw a scatter diagram of the eleven data points and visually check your conclusion.

ANSWERS

1. a) (-0.5823, 6.5823) (Did you remember to use 5 degrees of freedom?)
   b) Letting "1" refer to before and "2" refer to after the tranquilizer is given, we are testing H_0: $\mu_1 - \mu_2 = 0$ against the alternative H_a: $\mu_1 - \mu_2 > 0$ ($\mu_1 > \mu_2$). The null hypothesis is rejected whenever $t > 3.365$. Since the test statistic, t, equals 2.818, it does not fall in the rejection region. Thus, we do *not* reject H_0 and conclude that the tranquilizer does not significantly lower the pulse rate.

2. Notice that you can simply enter the new data into the STAT mode and it will append the original data. However, do not forget to change the number of degrees of freedom when finding values of t from the table.
   a) (0.7171, 4.7375) (Did you remember to use 10 degrees of freedom?)
   b) Letting "1" refer to before and "2" refer to after the tranquilizer is given, we are testing H_0: $\mu_1 - \mu_2 = 0$ against the alternative H_a: $\mu_1 - \mu_2 > 0$ ($\mu_1 > \mu_2$). The null hypothesis is rejected whenever $t > 2.764$. Since the test statistic, t, equals 3.7500, it does fall in the rejection region. Thus, we reject H_0 and conclude that the tranquilizer does significantly lower the pulse rate.
   c) Yes, the conclusion was different. The increased sample size gave more information about the population. Remember, though, that both of these results are from random sampling, not a census of the population.
   d) Notice that 9 of the 11 values (82%) are below the line. This visually indicates that the first coordinate (pulse rate before tranquilizer) values are more than the second coordinate values (pulse rate after tranquilizer). That is, it seems that the difference "pulse rate before tranquilizer - pulse rate after tranquilizer" is positive most of the time. The conclusion arrived at in c) seems correct.

INFERENCES ABOUT THE PROPORTIONS OF TWO POPULATIONS

The point estimator of the difference in the proportion of successes in two binomial experiments is the difference in the sample proportions of successes for independent random samples chosen from those populations. (See page 62 of this manual for the characteristics of a binomial experiment.)

LARGE-INDEPENDENT SAMPLES CONFIDENCE INTERVAL FOR $p_1 - p_2$

Use the following formula to give a (1-α)(100%) confidence interval for the difference in the binomial probabilities $p_1 - p_2$ of two populations when the following conditions hold true:

- the populations are binomially distributed
- the two samples are randomly and independently chosen from the respective populations
- it is appropriate to use the normal approximation to the binomial distribution; that is, the interval $\frac{x}{n} \pm 3\sqrt{\frac{(\frac{x}{n})(1-\frac{x}{n})}{n}}$ does not include 0 or 1 for *each* of the samples of size n_1 and n_2.

- The condition for when the normal approximation to the binomial distribution is "good" will vary among textbooks. Check your statistics text and modify the second condition accordingly.

PROGRAMMING

Select the AER-II mode. You should see on the display 33: TITLE?. Type in z [␣] c i [␣] p [SHIFT] 1 [−] p [SHIFT] 2 and press [ENT] to store the program name. Input the main routine at the M: prompt as follows:

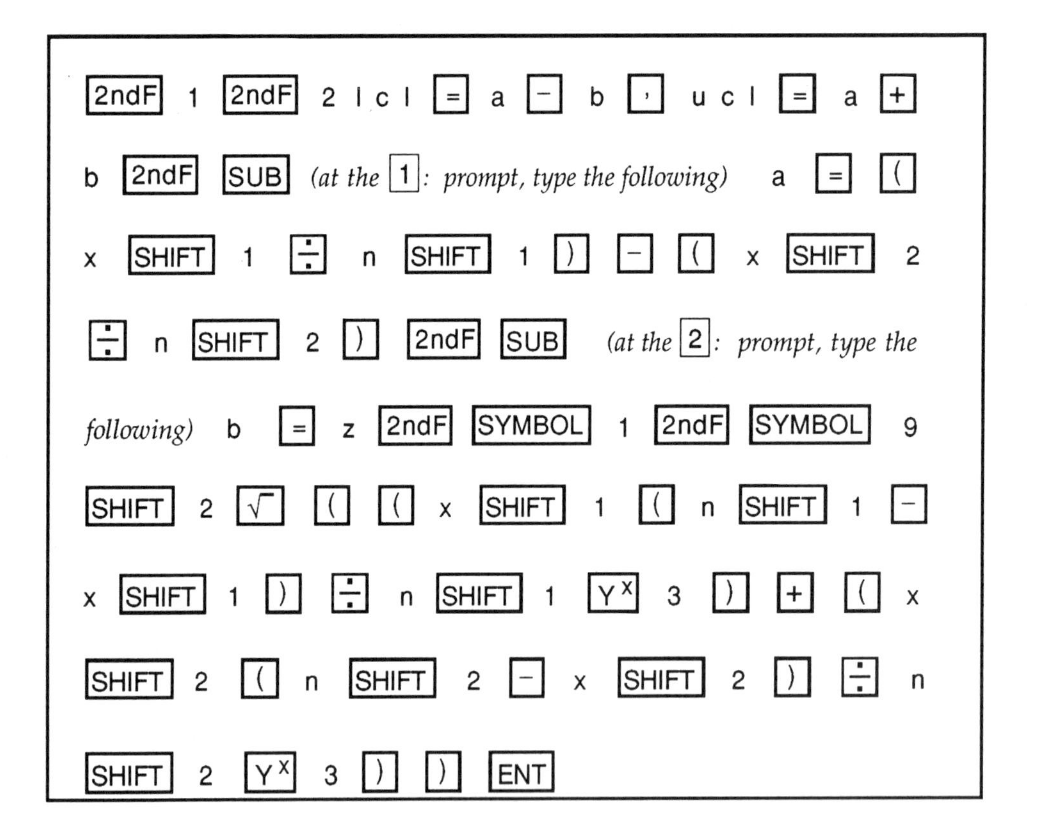

[2ndF] 1 [2ndF] 2 l c l [=] a [−] b [,] u c l [=] a [+] b [2ndF] [SUB] *(at the* [1]*: prompt, type the following)* a [=] [(] x [SHIFT] 1 [÷] n [SHIFT] 1 [)] [−] [(] x [SHIFT] 2 [÷] n [SHIFT] 2 [)] [2ndF] [SUB] *(at the* [2]*: prompt, type the following)* b [=] z [2ndF] [SYMBOL] 1 [2ndF] [SYMBOL] 9 [SHIFT] 2 [√] [(] [(] x [SHIFT] 1 [(] n [SHIFT] 1 [−] x [SHIFT] 1 [)] [÷] n [SHIFT] 1 [Yˣ] 3 [)] [+] [(] x [SHIFT] 2 [(] n [SHIFT] 2 [−] x [SHIFT] 2 [)] [÷] n [SHIFT] 2 [Yˣ] 3 [)] [)] [ENT]

The program " z c i p " is quite handy when you are checking the condition to see if it is appropriate to use the normal approximation to the binomial distribution. Execute this program in the COMP mode. When asked for the value of $z_{\alpha/2}$, enter the value 3. If the lower endpoint of either interval is negative, 0 is within the interval and this formula should not be used for the confidence interval estimating the difference in the binomial proportions. If the upper endpoint of either interval is greater than 1, then 1 is within the interval and this formula should not be used to give the confidence interval.

Once the program is entered, your calculator screen should look exactly like the following:

```
M: [1] [2] lcl=a-b,ucl=
a+b
```

Press [2ndF] [▽] to see the contents of subroutine 1 which should appear on your screen exactly like this:

```
[1]: a = (x1 ÷n1)-(x2 ÷
n2)
```

Press [2ndF] [▽] to see the contents of subroutine 2 which should appear on your screen exactly like this:

```
[2] : b = zα/2 √ ( (x1 (n1
-x1) ÷ n1 Y^x3) + (x2 (
n2 -x2) ÷ n2 Y^x 3) )
```

- Two subroutines are used here to avoid having to type in the same expression twice. Instead of displaying ANS1 and ANS2 as the respective lower and upper confidence limits of the interval, you will see "l c l " for the lower limit and "u c l " for the upper limit of the confidence interval.

EXERCISES

1. A random sample of 600 big-city residents, it was found that 485 own an automobile, but in a random sample of 400 small-town residents it was found that 375 own an automobile.
   a) Is it appropriate to use the normal approximations to the binomial distributions in this problem? Why or why not?
   b) Construct a 98% confidence interval for the actual difference in the proportions of all big-city residents and small-town residents who own an automobile.

2. The records of a certain hospital show that 65 men in a random sample of 100 men were admitted to the emergency room due to injuries suffered in automobile accidents versus 35 women in a random sample of 150 who were admitted to the emergency room due to injuries suffered in automobile accidents.
   a) Is it appropriate to use the normal approximations to the binomial distributions in this problem? Why or why not?
   b) Estimate the actual difference in the proportions of men and women admitted to this hospital due to injuries suffered in automobile accidents with a 90% confidence interval.
   c) Interpret this interval estimate.

ANSWERS

1. a) Yes, since the intervals (.5069, .7931) and (.9012, .9738) do not include 0 or 1.
   b) (-.1760, -.0824)

2. a) Yes, since the intervals (.7601, .8565) and (.1297, .3369) do not include 0 or 1.
   b) (.3198, .5135)
   c) We are estimating, with 90% confidence, that $p_m - p_w$ is in the interval (.3198, .5135). Thus, we are saying that p_m is estimated to be greater than p_w by an amount ranging anywhere from .3198 to .5135.

      If an interpretation on the confidence of 90% is desired, it is that out of many such intervals constructed in a similar manner, 90% of them will contain the actual difference $p_m - p_w$; 10% will not.

❐ LARGE-INDEPENDENT SAMPLES HYPOTHESIS TEST FOR $p_1 - p_2$

Tests of hypothesis involving the difference of two population proportions can be given for a hypothesized difference of zero or for a general difference $D \neq 0$. Because most applications call for a test of $p_1 - p_2 = 0$ (or equivalently $p_1 = p_2$), this is the situation considered in this manual. Consult your statistics text for the modifications to the formula to test $p_1 - p_2 = D$ for $D \neq 0$.

Use the following formula to give the value of the test statistic to be used in testing the hypothesis H_0: $p_1 - p_2 = 0$ against the alternative H_a: $p_1 - p_2 < 0$ (or $p_1 - p_2 > 0$ or $p_1 - p_2 \neq 0$) when the following conditions hold true:

- the populations are binomially distributed
- the two samples are randomly and independently chosen from the respective populations
- it is appropriate to use the normal approximation to the binomial distribution; that is, the interval $\frac{x}{n} \pm 3\sqrt{\frac{(\frac{x}{n})(1-\frac{x}{n})}{n}}$ does not include 0 or 1 for *each* of the samples of size n_1 and n_2.

PROGRAMMING

Select the AER-II mode. You should see on the display 34: TITLE?. Type in z [␣] h t [␣] p [SHIFT] 1 [−] p [SHIFT] 2 and press [ENT] to store the program name. Input the main routine at the M: prompt as follows:

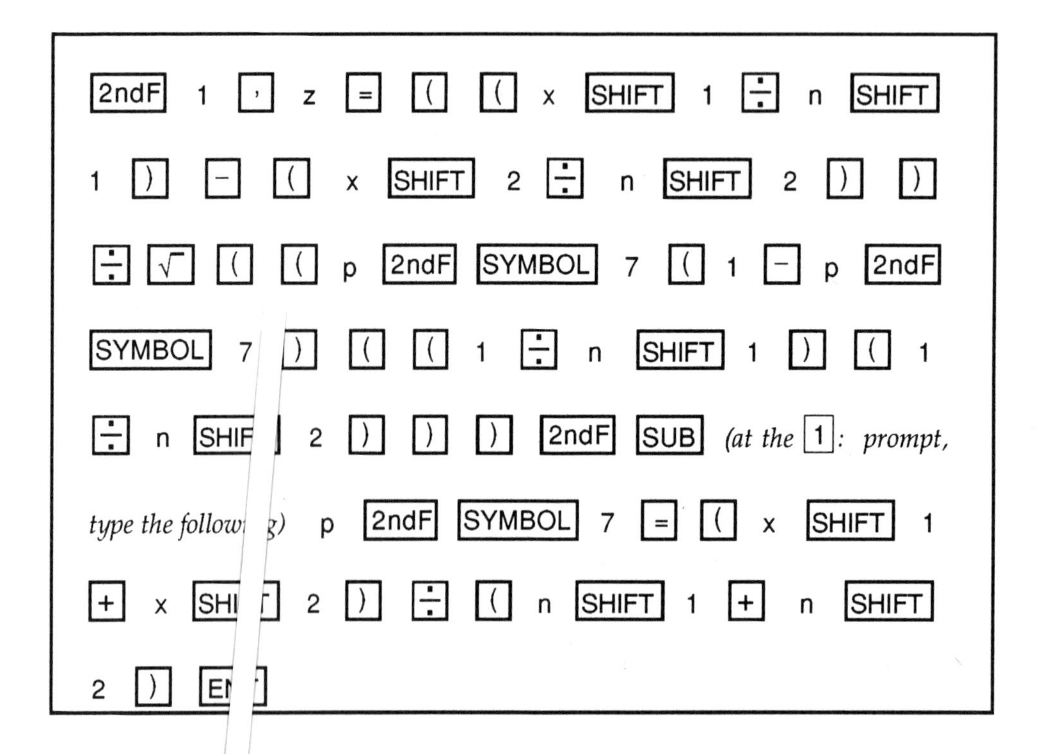

- Subroutin 1 calculates the value of the weighted average of the two sample proportions successes, x_1/n_1 and x_2/n_2. It is denoted on the calculator screen by p . This quantity is the weighted average of the two sample proportions successes with the larger sample receiving more weight.

Once the program is entered, your calculator screen should look exactly like the following:

```
M: [1] , z=( (x1 ÷ n1)−(
x2 ÷ n2) ) ÷ √ (p# (1− p
#) ( (1 ÷ n1) + (1÷ n2)
))
```

Press [2ndF] [▽] to see the contents of subroutine 1 which should appear on your screen exactly like this:

```
[1] : p# = (x1 +x2)÷(n1
+n2)
```

EXERCISES

1. A random sample of 350 men, ages 35 to 45, showed that 200 of them had life insurance. Another random sample of 465 men, ages 46 to 56, showed that 375 of them had life insurance.
   a) Is it appropriate to use the normal approximations to the binomial distributions in this problem? Why or why not?
   b) Does the sample evidence indicate, at the 0.01 level of significance, that the proportion of men in the age group 46 to 56 who have life insurance is greater than the proportion of men in the age group 35 to 45 who have life insurance?

2. Lee C. Charles is running for mayor. The local newspaper has said that the percentage of voters who favor Mr. Charles is different for the two precincts in town. Mr. Charles' staff randomly surveys a sample of the voters in each precinct and finds 79 voters out of 100 in favor of him in precinct 1 and 210 voters out of 300 favoring him in precinct 2. Is there evidence of a difference in the percentage of voters in favor of Mr. Charles in the two precincts at the 0.05 level of significance?

ANSWERS

1. a) Yes, since the intervals (.4921, .6508) and (.7515, .8614) do not include 0 or 1.
   b) $z = -7.2864$ so the decision is reject H_0; Yes (Note that p# = .7055.)

2. $z = 0.9980$ so the decision is do not reject H_0; No (Note that p# = .3510.)

CHAPTER 9

LINEAR REGRESSION

Many statistical problems are concerned with the relation, if such a relation exists, between two or more variables of interest. If the analysis involves two variables, we wish to know

- Are the two variables of interest related?
- If the variables are related, what is the nature of the relationship?
- If the variables are related in a linear form, how can information about one be used to predict the other?

When the values of the two variables are plotted, the resulting graph is called a *scatter diagram*. Many times it is the case that a possible relationship between the variables can be visually identified by looking at a scatter diagram of the data points.

The topic of *modeling* deals with trying to describe how variables are related. The linear regression model is a straight-line relation (fit) and enables us to estimate the value of one variable when it is related to another by a straight line. The estimated regression line, the straight line that "best fits" the data, is the line obtained through the least squares method. The estimated regression line is determined from the sample data and care should be taken when values outside the range of the sample data are estimated.

The general form of the population regression line is $y = \beta_o + \beta_1 x + \varepsilon$ where y is the dependent or response variable, x is the independent or predictor variable, ε is the random error component, β_o is the y-intercept of the line, and β_1 is the slope of the line. The equation of the line that relates the mean value E(y) to x is given by $E(y) = \mu = \beta_o + \beta_1 x$. The fitted line, calculated from the sample data points (x,y) is given by $\hat{y} = \hat{\beta}_0 + \hat{\beta}_1 x$ with the "hats" indicating that the values were calculated from sample data. Consult your statistics text for the model assumptions, that is, the conditions under which the regression model should be used, and for the formulas for the least squares estimates of the population y-intercept and slope.

■ TWO-VARIABLE STATISTICAL DATA

The formulas for the least squares estimates of the population parameters do not need to be programmed into the calculator because they are built-in functions in the STAT mode of the Sharp EL-5200. Let us first see how to enter two-variable data.

❐ ENTERING DATA

To enter data for linear regression (two-variable data), place the calculator in the data store mode of the STAT mode. To enter the data values (x,y), enter the x value, press the white [(x,y)] key, enter the y value and press the white [DATA] key. Continue in this manner until all data values are entered.

Consider the following: A manager of a small textile plant wishes to study the relationship between the time required to complete a certain task and the noise level at the work station. He collects data for a random sample of 5 employees and finds

Noise Level (x)	0.5	1	1.5	2	2.5
Time required to complete task (y)	1.3	1.8	2.9	3.6	4.8

(Assume, for purposes of illustration, that the noise level is rated on a scale of 1 to 5 and the time required to complete the task is given in minutes.) To enter this data, place the calculator in the STAT mode, choose data store, clear memory array S with the key sequence [TITLE / DATA] [2ndF] [CL] [ENT] [2ndF] [T-G-D] and press

.5	[(x,y)]	1.3	[DATA]
1	[(x,y)]	1.8	[DATA]
1.5	[(x,y)]	2.9	[DATA]
2	[(x,y)]	3.6	[DATA]
2.5	[(x,y)]	4.8	[DATA]

Your screen will appear as

```
.5,1.3 DATA
                    1.ØØØØ
1,1.8 DATA
                    2.ØØØØ
1.5, 2.9 DATA
                    3.ØØØØ
2,3.6 DATA
                    4.ØØØØ
2.5,4.8 DATA
                    5.ØØØØ
```

as the data values are entered. Recall that the numbers on the right of the screen give the number of data values that are entered.

DISPLAYING AND EDITING DATA

To display the data you have entered in memory location S, press TITLE/DATA and with the blinking cursor over the "S", press [=]. Data may be corrected at this point. If you wish to delete a data entry, with the blinking cursor over the position of the value to be deleted, press the [CD] key. If you wish to add more data, place the calculator in the text screen and follow the procedure for entering data values.

TRANSFER OF STATISTICAL DATA

If you wish to keep the data stored in memory location S for future use, it may be transferred to another unused memory location (for example, array C). Set the mode selector switch to COMP. Press [2ndF] [MATRIX (b)] to place the calculator in the MATRIX mode. Press [MAT] S [STO] [MAT] C and press TITLE/DATA .

To reverse the procedure and place the data back in memory S for use in the STAT mode, follow the same procedure as above replacing the letter S with the letter C and vice versa.

❒ WRITE PROTECT FUNCTION

If you are storing statistical data in another memory location for later use, it is a good idea to write protect the data so that it cannot be accidentally overwritten or erased. Place the calculator in the COMP mode and press [TITLE / DATA] key. Using the [▷] or [△] keys, position the blinking cursor to the letter corresponding to the array you wish to protect. As a result of pressing the [2ndF] and [PROTECT] keys, you will see the symbol "P" appear next to the dimension of the array.

To unprotect the data, move the cursor to the letter of the array where the write protection appears and press [2ndF] [PROTECT]. The symbol "P" will disappear and the write protection is removed.

- If you wish to name the data, have the blinking cursor over the letter of the array whose data you wish to give a name. Press the [▷] key and key in, using the letters on the right keyboard, the name of your data. Press [=] to store the name.

■ DRAWING THE SCATTER DIAGRAM AND REGRESSION LINE

Have the calculator in the data store mode of the STAT mode and with your data entered in memory location S, press [2ndF] [G(SD)] [AUTO] and [DRAW]. You will see on the screen the scatter plot of the data values. Since the Sharp El-5200 sets Xmin equal to the smallest x value and Xmax equal to the largest data value, these points are not very visible on the screen. To get a better picture, press [RANGE] and enter the following: Xmin = Ø, Xmax = 2.5, Xscl = Ø.5, Ymin = Ø, Ymax = 5, Yscl = 1. (Some of these settings may already be correct when you enter the range.) Exit the range and press 2ndF G(SD) DRAW . You should see the following on your display screen:

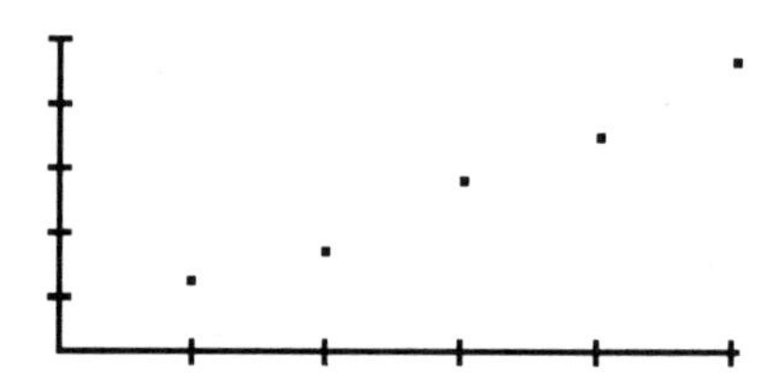

To superimpose the least squares regression line on your scatter plot, press 2ndF G(LR) DRAW .

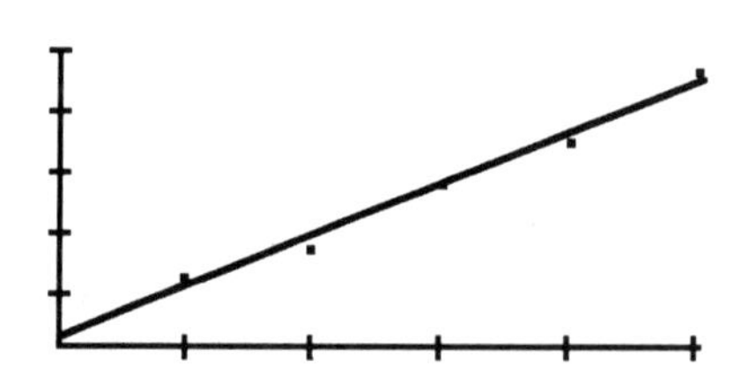

If you wish to view only the regression line, press 2ndF G(LR) AUTO DRAW . Note that keying in AUTO causes the previous graph to be erased. You may use the tracing function for values on the fitted line.

STATISTICS FOR TWO-VARIABLE DATA

You may obtain statistics for both x and y in the STAT mode after your two-variable data has been entered. The available statistics for both x and y are $\bar{x}$, $\bar{y}$, n, $\sum x$, $\sum y$, $\sum xy$, s_x, s_y, σ_x, σ_y, $\sum x^2$, and $\sum y^2$. These may be obtained on the text screen by pressing 2ndF and the appropriate key while in the STAT mode.

The estimates of the linear regression parameters may also be obtained from the calculator. For the fitted line, $y = a + bx$ (which may also be written as $\hat{y} = \hat{\beta}_0 + \hat{\beta}_1 x$) the slope b is obtained by pressing 2ndF MATRIX (b) keys. For our problem, b = 1.76ØØ. The y-intercept, a, is obtained by pressing 2ndF →DEC (a). For our problem, a = Ø.24ØØ. Thus, the estimated regression line computed from the sample data for the noise level problem is y = 1.76 + . 24x .

It certainly appears from the scatter diagram that the regression line gives a good fit to the data. However, you should perform a test of hypothesis on the slope of the line in order to see if x and y are really linearly related. Consult your statistics text for details. Now that you've seen a lot of programs, why not try yourself to program the formula for this hypothesis test?

Provided the fit is linear, we can use the regression line to obtain estimated values. To obtain the estimated value of x at a particular value of y, key in the value of y and press 2ndF →BIN (x′) . To obtain the estimated value of y at a particular value of x, key in the value of x and press 2ndF →OCT (y′) . For our example, suppose that the plant manager wished to predict the time required to complete the task at a noise level of 1.75. Since we wish to predict a value of y at a particular value of x, key in 1 . 75 2ndF →OCT (y′) to obtain the result 3.32

minutes. If the manager wanted to know the estimated noise level for someone who took a time of 4 minutes to complete the task, key in 4 [2ndF] [→BIN (x′)] to obtain 2.1364.

A measure of the strength of the linear relationship between x and y is the *correlation coefficient* , a quantity which expresses the extent to which two variables are related. The symbol for the sample correlation coefficient is r , while the symbol for the population correlation coefficient is ρ. Correlation coefficients will always range between -1 and 1. To interpret the sample correlation coefficient, we use the following:

- If $r = 1$, there is a perfect positive linear association between x and y with all the data points falling on a straight line with positive slope.

- If $r > 0$, there is a positive linear association between x and y that gets stronger as r gets closer to 1.

- If $r = 0$, there is no *linear* relationship between between x and y.

- If $r < 0$, there is a negative linear association between x and y that gets stronger as r gets closer to -1.

- If $r = -1$, there is a perfect negative linear association between x and y with all the data points falling on a straight line with negative slope.

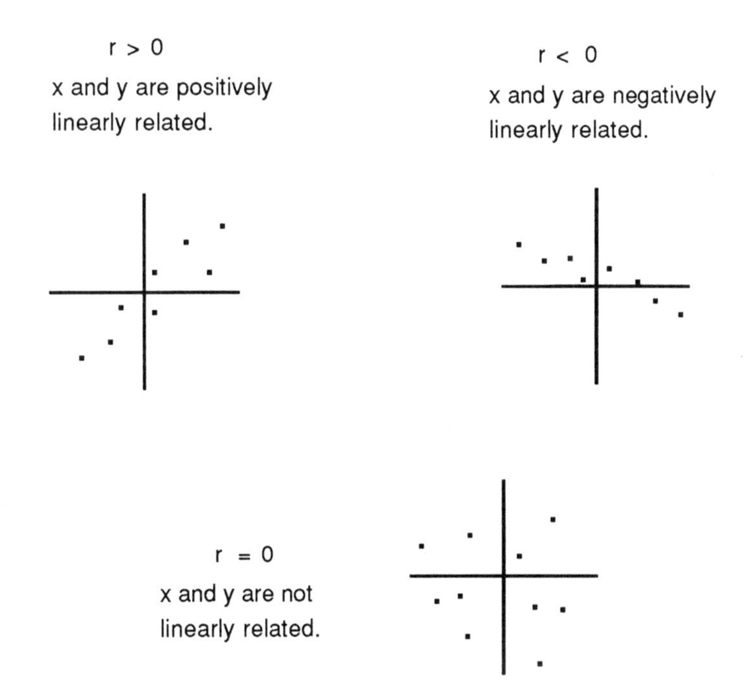

The sample correlation coefficient r may be obtained in the STAT mode by pressing 2ndF →HEX (r). For the noise level problem, r = .9921. Thus, assuming we have performed the hypothesis test showing that x and y are linearly related for this problem, it is evident that a very strong linear relationship exists between the time required to complete a particular task and the noise level in the room.

When interpreting the correlation coefficients, a word of warning is necessary. When values of r close to zero are found, it it tempting to conclude that there is little or no relationship between the variables under consideration. However, you must remember that the correlation coefficient measures only the linear relationship between the values. Even if r = 0, there could still be a relationship, possibly strong, between the variables, but it would be nonlinear. These methods measure the strength of a relationship; they do not prove that such a relationship exists or does not exist.

❐ THE MILE RUN

One interesting set of data to apply regression techniques to is the record for the mile run. The first runner in the twentieth century to break the record for the one-mile run was John Paul Jones of the USA in 1911. As obtained from the *World Almanac,* the table below shows the evolution of the record for the mile run from 1911 - 1980.

Record for the Mile Run
1911 - 1980

x (year)	Record	y (time in seconds)	
1911	John Paul Jones, USA	4:15.4	255.4
1913	John Paul Jones, USA	4:14.6	254.6
1915	Norman Taber, USA	4:12.6	252.6
1923	Paavo Nurmi, Finland	4:10.4	250.4
1931	Jules Ladoumegue, France	4:09.2	249.2
1933	Jack Lovelock, New Zeland	4:07.6	247.6
1934	Glen Cunningham, USA	4:06.8	246.8
1937	Sidney Wooderson, GB	4:06.4	246.4
1942	Gunder Haegg, Sweden	4:06.2	246.2
1942	Arne Andersson, Sweden	4:06.2	246.2
1942	Gunder Haegg, Sweden	4:04.6	244.6
1943	Arne Andersson, Sweden	4:02.6	242.6
1944	Arne Andersson, Sweden	4:01.6	241.6
1945	Gunder Haegg, Sweden	4:01.4	241.4
1954	Roger Bannister, GB	3:59.4	239.4
1954	John Landy, Australia	3:58.0	238.0
1957	Derek Ibbotson, GB	3:57.2	237.2
1958	Herb Elliott, Australia	3:54.5	234.5
1962	Peter Snell, New Zeland	3:54.4	234.4
1964	Peter Snell, New Zeland	3:54.1	234.1
1965	Michel Jazy, France	3:53.6	233.6
1966	Jim Ryun, USA	3:51.3	231.3
1967	Jim Ryun, USA	3:51.1	231.1
1975	Filbert Bayi, Tanzania	3:51.0	231.1
1975	John Walker, New Zeland	3:49.4	229.4
1979	Sebastian Coe, GB	3:49.0	229.0
1980	Steve Ovett, GB	3:48.8	228.8

Place your calculator in the STAT mode, clear memory array S, and enter this data for the mile run. Consider the year as the x value and the time in *seconds* as the y value. Thus, to enter the first data point, press 1 9 1 1 [(x,y)] 2 5 5 . 4 [DATA]

Continue in this manner until all 27 values are entered.

- You may want to save this data set for future use. If so, refer to the procedure for transferring data that is given on page 172 of this manual.

EXERCISES

1. Draw the scatter diagram for this data. Does it appear that the pattern is linear?

2. Draw the regression line for this data over the scatter diagram. Does it appear to "fit" the data?

3. Find the equation of the sample regression line for the 27 data points.

4. Even though is goes outside the range of the data, use the line (the appropriate calculator keys) to predict the record for the mile run in the years
   a) 1981
   b) 1985.

5. Enter into your calculator the actual records set for the years 1981 and 1985:

x (year)	Record	y (time in seconds)	
1981	Sebastian Coe, GB	3:48.5	228.5
1981	Steve Ovett, GB	3:48.4	228.4
1981	Sebastian Coe, GB	3:47.33	227.3
1985	Steve Cram, GB	3:46.31	226.3

   a) Use this data to predict the record for the mile run in the year 2000.
   b) When do you predict the three-minute mile will be broken?
   c) Find the equation of the sample regression line for the 31 data values.
   d) Explain how the slope of the regression line is affected by the addition of these four data values.

6. Calculate and interpret the correlation coefficient for the 1911-1985 data.

ANSWERS

1. Yes, you should find that the pattern appears linear.

2. Yes, it appears that the regression line visually gives a good fit to the data. You should, however, use the information in your statistics text to perform the necessary hypothesis test.

3. $y = 1037.4126 - 0.4089x$

4. a) 1981 227.3186
   b) 1985 225.7460
   (Compare your answers to these with the actual records given in the next problem. Comment.)

5. a) 2000 219.9455
   b) 180 seconds gives the year 2099 (2098.9682)
   c) $1027.1854 - 0.4036x$
   d) Notice that the slope has increased from the line calculated with the 27 data points. We would expect this to happen. If not, the line would eventually intersect the x-axis and we can't have someone running the mile in 0 seconds! As more data points are added, the line should fall less steeply.

6. $r = -0.9897$
   Since the correlation coefficient is negative, we see that as x (the year) increases, y (the time to run the mile) decreases. Also, since this value is so close to -1, we conclude that there is a strong linear relationship between the year in which the event takes place and the time required to run the mile.

CHAPTER 10

INTRODUCTION TO THE HP-28S, HP48SX AND HP48S

GETTING STARTED

The HP-28S, HP48S and HP48SX advanced scientific calculators are versatile tools with graphics and powerful computer algebra capabilities. The HP48S and HP48SX differ only in the expandable memory of the HP48SX, and operation of these two calculators is basically the same. These calculators are referenced in this manual as the HP48 calculators. There may be slight differences in programming between the HP-28S and HP48 calculators. When this is so, explanations and programs will be included in this manual for both calculators.

While many of the programs and techniques contained in this manual could be simplified and more efficiently programmed by experienced HP users, these notes are designed primarily to enhance a one-semester introductory statistics course with user understanding of the course material as well as the operation of the calculator.

Explanation of statistical concepts, simulation techniques and conditions for use of the formulas in your statistics course are explained in the first part of this manual. Even though the programs in the previous part of this manual are written for the Sharp EL-5200, Hewlett-Packard calculator users will find it most helpful to refer to the discussion of concepts that appears at the beginning of each section of this manual. These discussions are not calculator specific and will aid in your understanding of how you can use your calculator to enhance the subject matter.

Exercises appear at the end of most sections in Chapters 1 through 9 of this manual. Hewlett-Packard calculator users should work these exercises to verify that their programs are correctly entered.

Although it is assumed in this manual that Hewlett-Packard calculator users are already familiar with the basic operation of their machines, a few notes are included below for easy reference. Instructions and programs in the rest of this manual will be given first for the HP-28S and then for the HP-48 calculators. When no differences exist between instruction and programs for these calculators, the directions are not labeled for a specific calculator.

You will find it quite helpful to create a statistics directory to hold the programs in the remainder of this manual. For illustration purposes, this directory will be called ISTAT and specific directions for creating ISTAT and its associated subdirectories are given for each of the calculators in following sections.

■ HP-28S FUNDAMENTALS

If you are not familiar with the basic operation of the HP-28S, you will find it helpful to read Chapter 1 (Getting Started), Chapter 2 (Doing Arithmetic) and Appendix C (Notes for Algebraic Calculator Users) of the *Owner's Manual*[1]. If you wish an increased level of understanding into the theory and operations of the HP-28S, you are strongly advised to obtain the book *HP-28 Insights* by William C. Wickes[2].

❐ HP-28S ENTRY MODES

The HP-28S has three entry modes - Immediate, Algebraic and Alpha. Commands are executed in and arithmetic operations are performed in the Immediate entry mode. Algebraic objects are entered in the Algebraic entry mode while the Alpha entry mode is for entering programs.

---

[1]Hewlett-Packard HP-28S Advanced Scientific Calculator *Owner's Manual* and *Reference Manual*, Hewlett-Packard Company, Corvallis, Oregon, 1988.

[2]Wickes, William C., *HP-28 Insights: Principles and Programming of the HP-28C/S*, Larken Publications, Corvallis, Oregon, 1988.

When an object is entered in any of these modes, a blinking cursor appears in the command line. The HP-28S reminds you of the current entry mode with the style of the cursor. In the Immediate entry mode, the cursor is a blank box (▯), in the Algebraic mode, it is a box with two lines (▯), and in the Alpha entry mode, the cursor is a black box (▮).

❐ OPERATION PRINCIPLES OF THE HP-28S

The second-function (shift) key is the red key. Recall that it is used to activate the commands and menu keys printed in red above the keys on both keyboards. If it is pressed accidentally, simply press it again to deactivate the keystroke. If you make a mistake and an error message appears on the display, press [ON] to clear the error message. The numbers or commands that have previously been entered will remain on the stack display. The error message may also be cleared by pressing ■ [CLEAR] but previously entered values will be lost from the stack.

To tell your HP-28S how many decimal places are to be displayed, select the MODE menu by pressing ■ [MODE]. Key in the integer representing the number of decimal places you wish to use and press [FIX]ᴹ. The maximum number of digits shown on the HP-28S stack display is 12, and real numbers are shown with a range of 0 to 11 decimal places. (Internal calculations are always carried out to 15 digits regardless of the number of decimal places set by the user.) Numerical values exceeding the set number of decimal places are rounded in the normal manner.

Capital letters are obtained by pressing the corresponding key on the left keyboard. Lower case letters are obtained by pressing [LC] before pressing the desired letter key. Lowercase letters will continue until you press [LC] again to activate the capital letter mode.

Commands may be typed in using the left-keyboard alphabetic letters or with the proper menu key. To locate a particular menu key, press ■ [CATALOG] or consult the "Where" column in the HP-28S Operation Index on pages 323-349 of your *Reference Manual.*

Before you create your ISTAT directory, you should be in your HOME directory. Press [ON] and [△] *at the same time* and then press [USER]. You will now be in the HOME directory. Enter ['] I S T A T on the stack. Press ■[MEMORY] and [CRDIR]M. Press [USER] and you will see [ISTAT]M.

Press [ISTAT]M and you should see six blank menu boxes at the bottom of the display screen. It is a good idea to include a program that will let you return to your home directory when you desire. Enter the program QUIT by pressing the following keys: [<<] [HOME] [ENTER] ['] [QUIT] [STO].

■ HP48 FUNDAMENTALS

If you are not familiar with the basic operation of the truly unique HP48 calculator, you will find it helpful to read Chapter 1 (Getting Started) and Chapter 2 (The Keyboard and Display) in the *Owner's Manual*[1]. Readers who wish an increased level of understanding into the theory and operations of the HP48 are strongly advised to obtain the three volume *HP48 Insights* by William C. Wickes[2].

□ HP48 ENTRY MODES

The HP48 has four entry modes - Immediate, Algebraic, Program and Algebraic/ Program. Commands are executed in and arithmetic operations are performed in the Immediate (default) entry mode. Algebraic objects are entered in the Algebraic entry mode while the Program entry mode is used primarily for entering programs. The Algebraic/Program entry mode is used for keying algebraic objects into programs.

---

[1]Hewlett-Packard HP48SX Scientific Expandable *Owner's Manual,* Volumes I and II, Hewlett-Packard Company, Corvallis, Oregon, 1990.

[2]Wickes, William C., HP 48 *Insights,* Larken Publications, Corvallis, Oregon, 1991.

When an object is entered in any of these modes, a blinking cursor appears in the command line.

While the blinking cursor is the same (◆) in all these modes, the HP48 reminds you of the current entry mode with an annunciator that appears in the upper right-hand corner of the screen in all the modes except the Immediate entry mode.

❒ OPERATION PRINCIPLES OF THE HP48

There are two second-function (shift) keys on the HP48 single keyboard. Recall that the orange functions are obtained by pressing [◁] before the desired key and the blue functions are obtained by pressing [▷] before the desired key. If either of these keys are pressed accidentally, simply press the key again to deactivate the keystroke.

If you make a mistake and an error message appears on the display, press [ON] to clear the error message. The numbers or commands that have previously been entered will remain on the stack display. The error message may also be cleared by pressing [▷] [CLEAR] but previously entered values will be lost from the stack.

To tell your HP48 how many decimal places are to be displayed, select the MODE menu by pressing [◁] [MODES]. Key in the integer representing the number of decimal places you wish to use and press [FIX]^{M}. The maximum number of digits shown on the HP48 stack display is 12, and real numbers are shown with a range of 0 to 11 decimal places. (Internal calculations are always carried out to 15 digits regardless of the number of decimal places set by the user.) Numerical values exceeding the set number of decimal places are rounded in the normal manner.

Capital letters are obtained by first pressing [α] followed by the corresponding (white letter) key. The alpha (letter) entry mode may be locked by pressing [α] [α] and released by pressing [α]. (It is also released once the expression is entered onto the stack.) Lower case letters are obtained by pressing [α] [⬅] before pressing the desired letter key. When needed in writing a program, the lowercase letter entry may be locked by pressing [α] [⬅] [α] and will continue to be in effect until you press [α]. Special characters not printed on the keyboard may be accessed by pressing [α] [➡] followed by the proper key. A listing of alpha-right shifted keys is in the figure on page 52 of Volume I of your *Owner's Manual.*

Commands may be typed in using the alphabetic letters or with the proper menu key. To locate a particular menu key consult the "Description" column in the Operation Index on pages 707-821 of Volume II of your *Owner's Manual.*

To recall the contents of a variable or program to the stack, press [➡] followed by the menu key name of that variable or program. Remember, **R**ECALL = **R**IGHT shifted key ([➡]). To store or load what is in level one of the stack into a particular user menu item, say [ABC]M, press [⬅] [ABC]M This keystroke sequence is equivalent to ['] [ABC]M [STO]. Remember, **L**OAD = **L**EFT shifted key ([⬅]).

To create your ISTAT directory, enter ['] I S T A T on the stack. Press [➡] [HOME] [⬅] [MEMORY] and [CRDIR]M. Press [VAR] and you will see [ISTAT]M.

❏ EDITING

When entering keystrokes on the HP-28S, the previously entered commands are overwritten and insertions must be made using the [INS] key. When entering keystrokes on the HP48, the previously entered commands are inserted and deletions of unwanted commands must be made using the [DEL] key.

On both Hewlett-Packard calculators, if a mistake is realized before the command or numerical value is entered with the [ENTER] key, press the [←] key (backspace) to delete to the left from the cursor position and enter the proper keystrokes. Editing of values is possible after the [ENTER] key has been pressed through the use of the command line. With the information you wish to edit in line 1 of the stack, press [EDIT], key in the correct information and if necessary, delete the incorrect information, and press [ENTER].

To edit the contents of an existing variable or program, press ['] followed by the menu key containing the variable or program and then press [VISIT]. Press [ENTER] to install in memory the corrected program or variable.

❐ KEYS AND MENU SELECTION

Throughout this portion of the manual the following notation conventions will be used to help you recognize various commands and keystrokes:

- HP calculator keys will be enclosed in rectangular boxes except for numeric keys and decimal points (for example, [ENTER]).

- Menu keys for commands listed in various menus (whose names appear at the bottom of the lcd display screen) are identified by a small ^{M} to the upper right of the box (for example, [STEQ]^M).

- Menus usually include more keys than can be displayed across the bottom of the display screen. Press [NEXT] to scroll horizontally through the menu options. For convenience, these keys will not be shown in this manual.

• On the HP-28S, shifted keys are indicated by red type. These will be indicated by the (shifted key) name in a box preceded by a dark square to represent the red shift key (for example, ■ EDIT).

• On the HP48 calculators, shifted keys are indicated by either blue, orange or white type. These will be indicated by the (shifted key) name in a box preceded by ↱ to represent pressing the blue key to access a function in blue type, ↰ to represent pressing the orange key to access a function in orange type or α to represent pressing the alpha key to access an alphabetic letter in white type. These shifted symbols will not appear in programs in this manual.

• Note that some keys display on the screen different characters than what is printed on the keyboard:

KEY	DISPLAY
÷	/
1/x	INV( )
×	*
x^2	SQ( )
√x	√
0	Ø
y^x (HP48)	^
CHS (HP-28S)	±

CHAPTER
11

STATISTICAL GRAPHS ON THE HP-28S AND HP48

■ HISTOGRAMS

Statistical data on the Hewlett-Packard calculators is entered into a statistics matrix called ΣDAT. This matrix is the current statistical matrix and contains the data used by the commands in the STAT menu. If you wish to use data other than the current statistical matrix, you can choose another matrix by entering new data, editing the current data or selecting another matrix. If you wish to keep data you have entered for future reference, you can recall ΣDAT to level 1 of the stack and store the matrix under a name of your choice. Directions for entering single-variable and two-variable data are given in later sections of this manual.

You will find the calculator display easier to read for most problems in this section if you will choose the standard display mode for numbers by pressing ■ [MODE] [STD]M on the HP-28S or [↰] [MODE] [STD]M on the HP 48.

□ HISTOGRAM PROGRAMS FOR THE HP-28S

Before entering the following programs, you should create a subdirectory of ISTAT to contain the histogram programs. Press [USER]. If you see only [QUIT]M, you are already within your statistics directory. If you do not see [QUIT]M, press [USER] and then [ISTAT]M. Enter ['] H I S T on the stack. Press ■ [MEMORY] and [CRDIR]M. Press [USER] and you will see [HIST]M. You have just created a subdirectory of ISTAT called HIST. Press [HIST]M to enter your new directory and then include program QUIT as the first program.

The HP-28S has no built-in statistical graphing programs. However, courtesy of Bill Wickes of Hewlett-Packard Corporation, the programs below allow histograms to be drawn on the HP-28S. The program and subroutines for drawing the histogram should be placed in the directory HIST. Enter all programs before you try execution.

The histogram program and four associated subroutines that follow use a three-step process:

1. Accumulate data in the matrix ΣDAT.
2. Sort the data into equal width bins (intervals) using the program →BINS.
3. Plot the histogram with the program HIST which uses the vector of bins generated by →BINS. The bars' heights are originally equal in pixels to the count in the respective bins and are normalized with the subroutine FIT32 so that the largest element is 32. The bars are as wide as possible while fitting the requested number of bins into the display width of the calculator screen.

PROGRAMMING

Program: →BINS

Purpose: Sorts the contents of the matrix into equal width intervals (bins)

User Input: On the stack, input on level 3 the lower limit of the acceptable range, input on level 2 the upper limit of acceptable range, and on level 1 input the number of bins which equals number of intervals used.

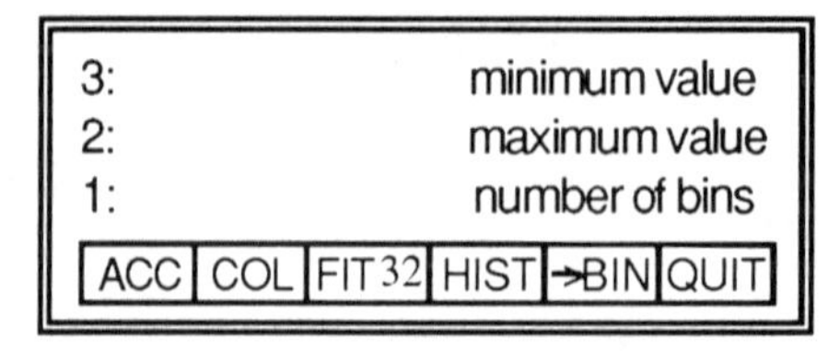

```
<< 3 ROLLD OVER - → n xmin diff
   << n 1 →LIST ∅ CON 1 NΣ
      FOR i 'IP(n*(ΣDAT(i,1) - xmin)/diff+1)' EVAL
      DUP DUP
      IF ∅ > SWAP n ≤ AND
      THEN DUP2 GET 1 + PUT
      ELSE DROP
      END
      NEXT >>
>> [ENTER] ['] ■[→] B I N S [STO]
```

Program: HIST

Purpose: Plots the elements of the →BINS vector as a histogram. All elements must be non-negative and the number of bins must be fewer 69.

User Input: None

```
<< FIT32 CLLCD DUP SIZE LIST→ DROP
   137 OVER / IP 1 - → d n w
   << "" DUP DUP2 1 n
      FOR m d m GET COL
      w ACC 4 DUPN + + + →LCD
      NEXT 137 n w 1 + * - 1 -
      IF DUP THEN 255 CHR ELSE ""
      END DUP DUP2 5 ROLL ACC
      + + + NOT →LCD >>
>> [ENTER] ['] H I S T [STO]
```

Program: FIT32

Purpose: Scales entries in the →BINS vector so that the largest is 32

User Input: None

```
<< DUP 1 GET SWAP DUP SIZE LIST→
   DROP 1 DUP ROT START GETI 4 ROLL
   MAX 3 ROLLD NEXT DROP SWAP / 32 *
>> [ENTER] ['] F I T 3 2 [STO]
```

Subroutine: COL

Purpose: Creates four 1-character strings that represent a pixel column whose length equals n, the argument to COL, where $0 \le n \le 32$.

User Input: None

```
<< 32 SWAP - IF DUP THEN STWS # Ød 1 - 32
   STWS 1 4 START DUP #FF h AND B→R CHR
   SWAP RRB NEXT DROP ELSE CHR DUP DUP2 END
>> [ENTER] ['] C O L [STO]
```

Subroutine: ACC

Purpose: Concatenate n copies of stra - strd to strA t - strD, respectively, then concatenate chr(255) to each.

User Input: None

```
<< → n
   << 1 4 START 5 ROLL 1 n START OVER +
      NEXT 255 CHR + 8 ROLLD 4 ROLLD NEXT
      4 DROPN >>
>> [ENTER] ['] A C C [STO]
```

❑ ENTERING AND EDITING ONE-VARIABLE DATA ON THE HP-28S

Data should be entered into ΣDAT using the STAT menu on the calculator before activating these programs. Follow the steps below to enter single-variable data on the HP-28S:

- Enter the STAT mode by pressing ■ [STAT].
- Clear any previous statistical data by pressing [CLΣ]M.
- Enter the data values, one by one, pressing [Σ+]M after each entry.
- When you finish entering the data values, press [NΣ]M to check that the correct number of data values has been entered.
- Press [USER] and you will see as the newly created statistical data matrix listed in the menu as ΣDAT.

If you make a mistake while keying in the data, you can correct the mistake before you press [Σ+]M by pressing [◀] and entering the correct value. To verify that the correct values have been entered or to correct an incorrect value that has already been entered, key in ['] [ΣDAT]M ■ [VISIT]. Use the four cursor keys ([▽], [△], [◁] and [▷]) to scroll through the data and edit if necessary. Press [ENTER] to store the edited ΣDAT matrix.

- If you wish to keep this data for future reference, key in ['] NAME [STO] where "NAME" is a description of your choice. Press [NAME]M to recall the data to the stack.

- If you wish, you may refer to your *Owner's Manual* for entering statistical data as a matrix and/or other methods of editing data.

DRAWING THE HISTOGRAM

Refer to the gasoline data that is given on page 17 of this manual. Enter the data into your calculator. To draw the histogram, you must first use the →BINS program to obtain the frequency count in the classes (intervals or bins) of the histogram. Return to the STAT menu and press MINΣ M and MAXΣ M to enter these values on the stack. Next enter the number of classes (intervals or bins) you wish for your graph. For illustration purposes, let's enter 5.

Return to the USER menu and press →BIN M. You will see [3 6 5 4 1] on the stack. These values represent, in order from left to right, the frequencies of the classes of the frequency distribution for this data. (Refer to page 22 of this manual for the resulting frequency distribution.) The (approximate) width of each interval is given by the formula

$$\text{width of class (bin)} = \frac{\max\Sigma - \min\Sigma}{\text{the number of classes}}.$$

For this data, with 5 classes, the width of each class is 3.4. Unfortunately, there is a problem. There were originally 20 data values and the sum of the frequencies in the BINS vector is only 19. Why? It is because the calculator will not include a value falling on the right boundary of a class interval in that interval.

Let's repeat the above procedure, but this time, add 0.5 to MAXΣ. After executing →BIN M you should see [3 6 5 4 2] on the stack. Notice that all data values have now been included.

- You will find it helpful to store the results of the program →BINS. After executing this program, the results will be on level 1 of the stack. Press ' B STO and then press B M to replace the results on the stack for use by the program HIST. Notice that you now have two additional items in your menu, B created by your process of storing the results of →BINS and ΣDAT created by your original entry of the data. (There will be a third new item if you have stored your data under a name of your choice.)

To draw the histogram, press [HIST]M.

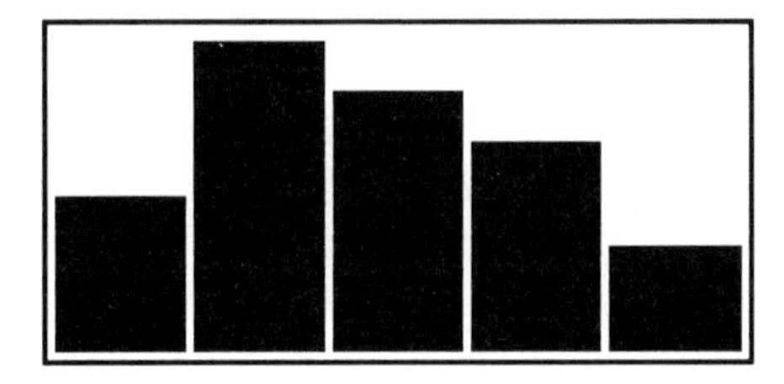

- If you wish to see the effect of the subroutine FIT32 , eliminate "FIT32" from the first line of the HIST program. Recall B, your variable containing the results of the program →BINS to level 1 of the stack, press [FIT32]M, and draw the histogram.

You will probably want to reinsert FIT32 at the beginning of your histogram program.

There may be times when you wish a particular histogram saved for future reference. The following program, GETG (get graph), will recall the elements of the graphics screen that have been saved in the variable named SCR (save screen).

<< SCR →LCD DGTIZ >> [ENTER] ['] GETG [STO]

To store the contents of the graphics screen in variable SCR, draw the histogram, press [ON] to return to the stack, press ■ [LAST] to recall to the stack the graph data "■■■■■■■■" and press ['] S C R [STO]. You may easily move this to another variable name by recalling it to the screen with [SCR]M and storing it in the name of that new variable. To recall a particular graph, however, it must be restored under the name SCR. Note that the cursor is active when program GETG is used to recall a graph.

❐ HISTOGRAMS ON THE HP48

The HP48 has two built-in programs for drawing histograms. These are included in the STAT menu are are called BARPL and HISTP. The autoscaled barplot command BARPL on the HP48 is unique in that it plots bars of both positive and negative heights. The autoscaled histogram plotting command HISTP has the unique feature of plotting a histogram of *relative* frequencies.

In order to obtain control over the number of rectangles, it is relatively easy to use the command BINS to obtain an interval frequency count and then use BARPL to obtain a histogram of numerical frequencies. Masking of values less than the minimum x-value or greater than the maximum x-value is accomplished with the BINS command in the STAT menu. You can also, with a single keystroke, reset the value of the resolution in order to obtain a specified number of intervals when using the HISTP command to construct the histogram.

Histograms may also be drawn with the user specifying the x and y ranges in the PLOT menu and choosing either BAR or HIST as the plot type. However, before the histogram can be drawn, data must be entered in the calculator in the statistical data matrix ΣDAT.

❑ ENTERING AND EDITING ONE-VARIABLE DATA ON THE HP48

The statistical data matrix ΣDAT you create will reside in the active subdirectory of the VAR menu. If you wish to use this data when you change to another subdirectory, you must store the matrix in the subdirectory to which you have changed. Different ΣDAT matrices may exist in various places in your calculator. You should always recall ΣDAT to the stack to be sure you are working with the correct data if you have changed directories and have not renamed ΣDAT. The instructions below are for entering data directly into the STAT menu using the instructions that appear at the top of the display screen. Note that you can also use the MatrixWriter application to enter statistical data and store it as the current statistics matrix using [NEW]M. Instructions for that procedure are on page 370 of Volume I of your *Owner's Manual*.

Follow the steps below to enter single-variable data on the HP48:

- Press [VAR] and enter the subdirectory in which you wish to have the data stored.
- Enter the STAT mode by pressing [◁] [STAT].
- Clear any previous statistical data by pressing [CLΣ]M.
- Enter the data values, one by one, pressing [Σ+]M after each entry.
- Notice that as you enter the data values the last value entered appears at the top of the display screen. When you finish entering the data values check the number in the parentheses following ΣDAT to verify that the correct number of values have been entered. (The number of entered data values can also be obtained by pressing [NΣ]M.)
- Press [VAR] and you will see as the newly created statistical data matrix listed in the menu as ΣDAT.

If you make a mistake while keying in the data, you can correct the mistake before you press [Σ+]M by pressing [◆] and entering the correct value. To verify that the correct values have been entered or to correct an incorrect value that has already been entered with [Σ+]M, press [EDITΣ]M. Use the cursor keys ([▽] and [△]) to scroll through the data and edit if necessary. Press [ENTER] to store the edited value and [ENTER] again to store the edited ΣDAT matrix.

- If you wish to keep this data for future reference, key in ['] [α] N A M E [STO] where "NAME" is a description of your choice. This matrix is then labeled and copied into the STAT catalog. To recall that matrix to the stack , press [CAT]M, move the pointer to the name of that matrix, and press [→STK]M [ON] or [ENTER] [ON].
- If you have previously entered data, stored it under a name of your choice and wish to recall the data to the STAT menu, follow the above procedure to recall the data to the stack. Press [STOΣ]M to make this matrix the current statistical matrix.

DRAWING THE HISTOGRAM

Let's briefly explore the basic difference in the two procedures for drawing histograms. Press [◁] [STAT] and if you do not see the message "No current data. Enter data point, press Σ+", press [CLΣ]M to clear the current statistical matrix. (If you do not clear the data that was previously entered, your new data will be appended to that already in the matrix.) Enter the nine data values

1 2 3 4 5 4 3 2 1

Find [BARPL]M in the STAT menu and press the key. You should see the graph

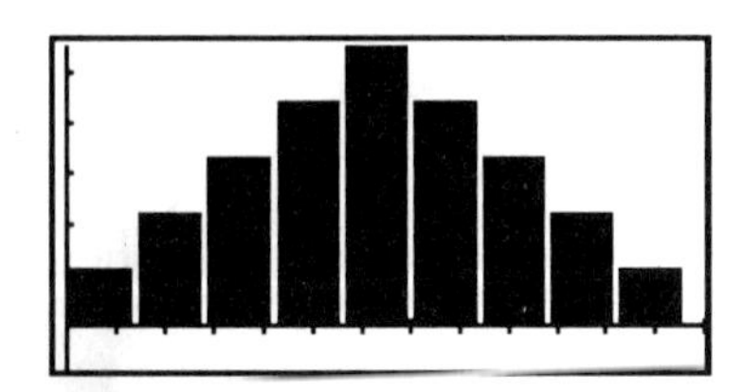

Notice that there are 9 bars (intervals) with the *height* of each bar shown as the data value itself. Press [↱] [PLOT] and you will see that the x range is 0 to 9 and the y range is -.75 to 5. Since there are 9 rectangles and the width of the screen is 9, the width of each rectangle is approximately 1 unit (disregarding the small space between the bars). (Why do you think the ymin value is negative?) Next, find [HISTP]^{M} in the STAT menu and press the key. You should see the graph

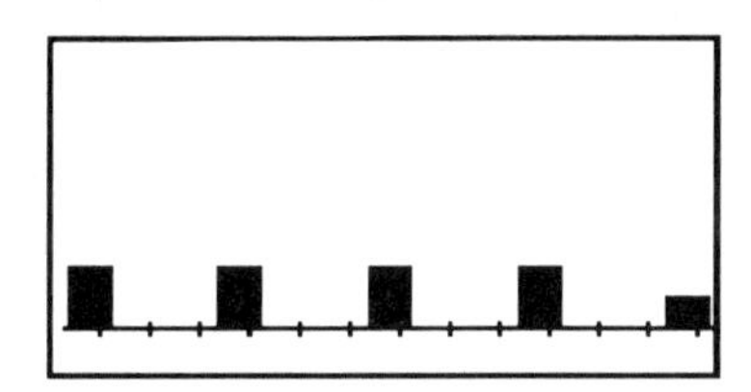

Notice that even though you only see 5 bars, there are actually 13 invervals on the horizontal axis that are being shown. Press [↱] [PLOT] and you will see that the x range is 1 to 5 and the y range is -1.35 to 9. Since there are 13 intervals and the width of the screen is 4, the width of each rectangle is a decimal approximation of 4/13. Since there are 2 of the data values between 1 and 17/13, namely the two ones, the height or frequency of the first rectangle is 2. You should verify that the fourth interval contains the two two's, etc. Thus, with [HISTP]^{M}, the data values have been grouped with the height of each rectangle being the number of data values in the particular interval.

Refer to the gasoline data that is given on page 17 of this manual. Enter the data into your calculator. (If you wish to obtain the same graphs as the ones to follow, enter the data as read down the columns.) You may use the name supplied by the calculator or rename the matrix GAS. Press [BARPL]ᴹ and you will see the following graph composed of 19 rectangles.

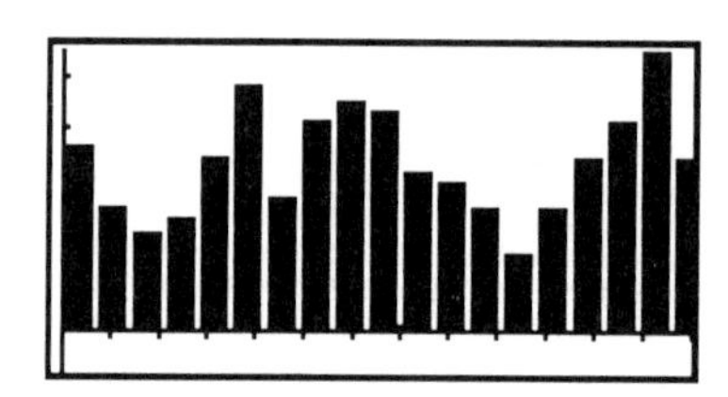

If you look closely at this graph, you will see that the first 19 data values that were entered have been graphed with the data values as the heights of each of the rectangles.

- Notice that when the graph is drawn on the screen, the PLOTR menu is active and you can estimate the heights of the rectangles with the cross-hair cursor and [COORD]ᴹ. Realize, however, that the coordinates appearing at the bottom of the screen are pixel coordinates, not points on the graph you have drawn.

- The cross-hair cursor is easily lost with all these rectangles on the screen. To quickly move this cursor to the upper left-hand corner of the screen, press [↱] [◁] [↱] [△] and to quickly move it to the upper right-hand corner of the screen, press [↱] [▷] [↱] [△].

- If you do not wish to have the menu keys at the bottom of the screen when viewing the graph, press [−]. If you do not wish to have the menu keys but you do want to see the cursor coordinates, press [+]. Pressing the key once again will return the menu keys to the graphical display.

Press [↱] [PLOT] and notice the x and y range settings. We entered 20 data values yet there were only 19 rectangles. Why? The calculator did not include the value falling on the right boundary of a class in that interval, and the autoscaling used in the BARPL command sets the maximum value of x to the largest data value. Also notice that this graph would be different if you had entered the data values in a different order. Obviously this graph is not the histogram we wish to draw. Return to the STAT mode and let's see what happens when we press [HISTP]^{M}:

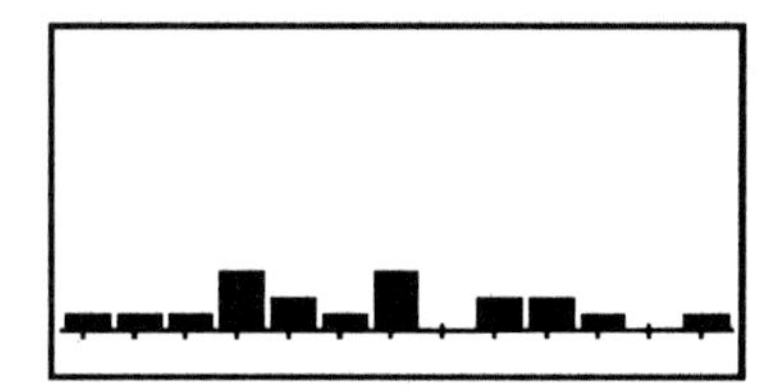

Press [ON] to return to the stack and key in [↱] [PLOT]. Notice the x and y range settings. As you see, the minimum value of the x range is set at the smallest data value, 6, and the maximum value of the x range is set at the largest data value, 23. HISTPLOT uses autoscaling and will always construct a histogram of 13 rectangles (classes) unless you change the default resolution setting. The (approximate) width of each interval is given by the formula

$$\text{width of class} = \frac{\text{max}\Sigma - \text{min}\Sigma}{\text{the number of classes}}$$

For this example, the class width is $(23 - 6)/13 \approx 1.31$. Since the first class starts at 6, the last class ends at 23.03. Note that the maximum value of 23 is included in this last class, but is not on the boundary, and was therefore counted in the frequency distribution and graphed.

Neither of the graphs drawn by pressing the BARPLOT or HISTPLOT keys is the "best-looking" histogram for this data. There are, however, two methods of constructing a histogram on your HP48 that allow you to select the number of class intervals. Before we begin this example, store the data you have in the current statistical matrix as GAS (See page 198 of this manual for the procedure).

One method for the construction of a histogram with a specified number of intervals is to use BINS command (in the STAT menu) in conjunction with BARPLOT. The BINS command allows you to see the frequencies of the classes *before* the graph is drawn. To use this command, you must follow these steps:

1. Enter, on the stack, the minimum x-value to use (this is normally the smallest data value).
2. Enter, on the stack, the width of each bin (class) that you determine by the formula on the previous page of this manual.
3. Enter, on the stack, the number of bins (classes) you wish to use.
4. Press [BINS] M.

The output to the stack of the BINS command is

Level 2: A "bins" matrix containing the frequencies of the classes.

Level 1: An "excess" vector containing two numbers. The first number is the number of data values less than the minimum x-value you specified and the second number is the number of data values greater than the maximum x-value. This vector in level 1 should be [0,0] if you wish all data values included in your histogram.

To draw the histogram,

1. Drop the excess vector from level 1 of the stack with [⬅] [DROP].
2. Make the bins vector the new ΣDAT matrix by pressing [STOΣ] M. *This will delete the original data unless you have saved it under a different name.*

3. Press BARPL^{M}.

- Notice that the x-range settings and cross-hair cursor coordinates are now in terms of the number of bins, not the original data.

Let's construct a histogram for the gasoline data you have entered using the above procedure. Be sure you have this data as the current statistical data. Let's choose to draw a histogram of 5 classes. This gives the class width as (23-6)/5 = 3.4. Enter on the stack in level 1 the minimum data value of 6, the class width of 3.4 on level 2, and the number of bins, 5, on level 3. Press BINS^{M} and see

Press ⬑ DROP to drop the excess vector showing that all data values have been included, and press STOΣ^{M} and BARPL^{M} to obtain the histogram

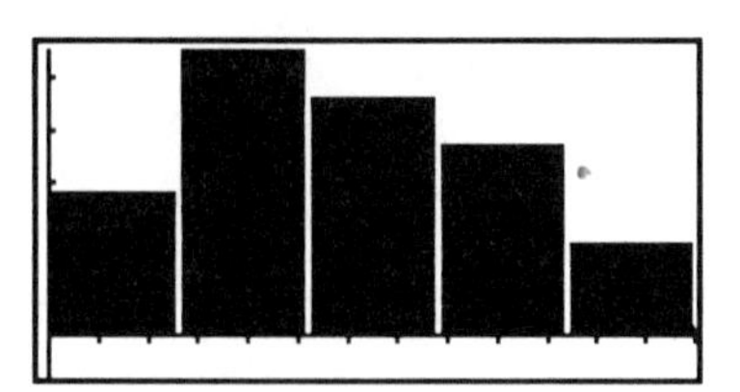

A second method for construction of your histogram is to use the RES command in the PLOT menu. This command allows you to change the number of classes by changing their width. The default setting for RES is 0. If you wish to see the value for the resolution setting on your calculator, press ⬏ RES^{M}. To change the resolution, enter the new value either as a real number or a binary integer argument and press RES M to store that value. For HISTOGRAM plot types, RES determines the width of each class into which the data is placed. (Be sure to press 0 RES M or RESET M after experimenting with different values for RES to restore the default setting.) Remember that the width of each interval (not counting for the small space on the graph between the rectangles) is determined by the formula

$$\text{width of class} = \frac{\max\Sigma - \min\Sigma}{\text{the number of classes}} \quad \text{(rounded up if you round)}$$

The maximum value of 23 minus the minimum value of 6 divided by the number of classes, say 5, gives 3.4. Enter 3 . 4 and press RES M.

Recall that you do not have the original data stored in the current statistical data matrix since you stored the matrix of bins in the previous example. Press GAS M (located in your VAR menu) and store it as the current statistical matrix with STOΣ M. Press ⬅ PLOT PTYPE M HIST M 6 SPC 23 XRNG M ERASE M DRAW M . If you do not get the same graph as the one on the previous page, reset the y range to appropriate values (for instance, -1 to 6), and press ERASE M DRAW M to redraw the histogram.

Experiment with different values for RES and redraw this histogram. What happens to the width of the intervals as RES increases from 3.4? What do you notice as RES decreases from 3.4? Do not forget to reset the resolution to the default value of zero when you have finished with the histogram procedure.

❒ STORING AND RECALLING GRAPHS ON THE HP48

There may be times when you wish a particular histogram saved for future reference. You may store the histograms you wish to keep in a graphing directory. This will allow storage of many graphs without cluttering your ISTAT directory. If you wish to create a subdirectory of ISTAT, say GRPH, to contain your graphs, press the following keys: [VAR] [↱] [HOME] [ISTAT]M. Enter ['] G R P H on the stack and press [↰] [MEMORY] and [CRDIR]M. Press [VAR] and you will see [GRPH]M. You should press [GRPH]M to enter your new directory before storing any graphs and include any of the following programs you wish to use in this directory.

To *store a graph* that you have just constructed on the calculator, press [◁] to return the graph to the screen. Press [ON] [STO] to store the graph as a graphic object and see "Graphic 131 x 64" on level 1 of the stack. Store this in a name of your choice, let's say GASH (gas histogram), with ['] G A S H [STO]. Press [VAR] and you will see that GASH has been listed in the user menu of your GRPH directory.

- You should not use the name PICT, PVIEW or P for your graph as these are reserved for programs or names within the calculator memory.

There are several methods you may use to *recall a graphic object* that you have stored.

1) Pressing [◁] will recall the most recently drawn graph.

2) If you only desire to see the graph you have stored and not have the cursor or the graphic environment active, press the menu key of the variable in which you have stored the object (here [GASH]M) which returns "Graphic 131 x 64" to level 1 of the stack and then press [PRG] [DSPL]M [→LCD]M.

3) Program SEE may be used after the graph has been drawn and stored to recall the graph to the screen. The cursor and menu keys in the graphing environment are active after using program SEE. If you recall a graph using SEE after the range has been changed to draw another graph, the coordinates you see at the bottom of your screen will refer to points in the range that is now active in the calculator, not the range in which you originally had drawn your graph.

The two versions of program SEE that follow accomplish the same purpose. Both are included for you, but *you should choose only one of the two* to place in your calculator. Version 1 is recommended.

The following program should be entered in the GRPH directory and then stored with the keystrokes ['] SEE [STO] . (Program SEE will work in any directory - it is more convenient to have it in the directory in which you have stored your graphs.) To use version 1 of this program, you must first recall the graph you wish to view to level 1 of the stack, execute the program, and then press [◁]. For instance, to "see" GASH, you would press [GASH]M, press [SEE]M and then [◁].

VERSION 1 PROGRAMMING

```
<<  ERASE  PICT  STO  >>
```

Before using version 2 of this program, you must first recall the graph you wish to view to the stack and store it in P. For instance, to "see" GASH, you would enter [GASH]M ['] P [STO] and then execute program SEE.

VERSION 2 PROGRAMMING

```
<<  ERASE  {  #  Ød  #  Ød  }  PVIEW  {
#  Ød  #  Ød  }  PICT  OVER  P  GXOR
GRAPH  DROP  >>
```

EXERCISES

The Tobacco Institute[1] reports the following list of state taxes (per pack) placed on cigarettes.

Ala.	16.5¢	Mont.	18.0¢
Alaska	29.0¢	Neb.	27.0¢
Ariz.	18.0¢	Nev.	35.0¢
Ark.	22.0¢	N. H.	25.0¢
Calif.	35.0¢	N. J.	40.0¢
Colo.	20.0¢	N. M.	15.0¢
Conn.	40.0¢	N. Y.	39.0¢
Del.	24.0¢	N. C.	2.0¢
D. C.	30.0¢	N. D.	30.0¢
Fla.	33.9¢	Ohio	18.0¢
Ga.	12.0¢	Okla.	23.0¢
Hawaii	42.0¢	Ore.	28.0¢
Idaho	18.0¢	Pa.	18.0¢
Ill.	30.0¢	R.I.	37.0¢
Ind.	15.5¢	S. C.	7.0¢
Iowa	36.0¢	S. D.	23.0¢
Kan.	24.0¢	Tenn.	13.0¢
Ky.	3.0¢	Texas	41.0¢
La.	20.0¢	Utah	26.5¢
Maine	33.0¢	Vt.	18.0¢
Md.	16.0¢	Va.	2.5¢
Mass.	26.0¢	Wash.	34.0¢
Mich.	25.0¢	W. Va.	17.0¢
Minn.	43.0¢	Wisc.	30.0¢
Miss.	18.0¢	Wyo.	12.0¢
Mo.	13.0¢		

Enter this data in the STAT mode of your calculator and find the following:

1) the mean state tax placed on a pack of cigarettes
2) the standard deviation of the state taxes
3) a histogram of this data consisting of 8 class intervals of equal width. What is the width of each interval? What skewness is shown? What are the interval frequencies?

---

[1] *USA Today*, June 21, 1991, Source: The Tobacco Institute

ANSWERS

1) The mean of 23.96¢ per pack is obtained by pressing [MEAN]M in the STAT menu.

2) The standard deviation of 10.55¢ is obtained by pressing [SDEV]M in the STAT menu.

3) The maximum and minimum values are, respectively, 43¢ and 2¢. Dividing the difference in these by 8, the number of classes, we obtain 5.125 which is stored in [RES]M. By choosing the plot type [HIST]M, setting the x range from 2 to 43, the y range from -1 to 11 and pressing [ERASE]M [DRAW]M we obtain the histogram

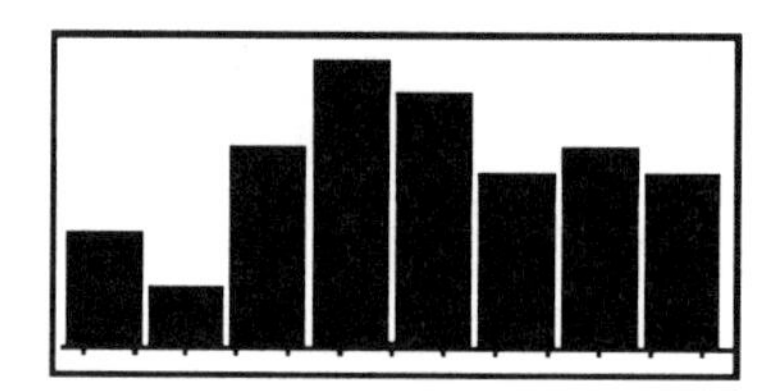

The frequencies of the intervals in this graph can be obtained by using the arrow keys on the calculator to move the crosshair cursor to the top of each of the rectangles and obtaining a pixel coordinate estimate of the height. That value should be rounded to the nearest whole number. This procedure may be tedious, however.

Note that you could also have drawn this histogram using the BINS and BARPLOT commands. This method of constructing the histogram will give you a vector of frequencies without having to estimate them from the graph. First enter 2, the minimum data value, on the stack, then enter 5.125, the width of each class, and then enter 8, the number of intervals desired. Press [BINS]M, drop the excess vector [0,0] showing that all the data values have been included, [ENTER] to make a copy of the frequencies so that you can store the bins vector for later recall, [STOΣ]M and [BARPL]M. You will see the same histogram given above. The bins vector containing the frequencies is [4 2 7 10 9 6 7 6].

PERCENTILES

When you describe the position in relation to the other measurements of a particular measurement in a data set, you are using a measure called a *percentile*. When your data set is arranged in order from smallest to largest, the pth percentile is a number (which may or may not be one of the data values) such that p% of the measurements fall at or below that number. The median of a data set is the 50th percentile.

The 25th percentile is called the *lower quartile*, Q_1, and is the median of the lower half of the data. The 75th percentile is called the *upper quartile*, Q_3, and is the median of the upper half of the data. The *interquartile range* is the difference $Q_3 - Q_1$ and tells us the spread of the middle half of the data.

Measurements must be in order before you can find percentiles and your HP calculator does not have a built-in sorting routine. However, if you enter the data onto the stack in any order and use program TLST to form a list containing the data, program SORT places the data in that list in ascending order so that the percentiles can be calculated from within the box plot program. The sorted list is stored in the menu under the name DATA.

SORT AND PERCENTILE PROGRAMS FOR THE HP-28S

Because there are several programs used to find percentiles, it is best to create a directory containing these programs. Enter your statistics directory by pressing [ISTAT]^M and press [MEMORY] ['] PCNT [CRDIR]^M to create a subdirectory called PCNT. To see that this directory has been included in your user menu, press [USER] and look at the menu. Press [PCNT]^M to enter the PCNT directory before entering the remaining programs in this section. You should include program [QUIT] as the first program entered in this new directory.

The following program should be entered in the PCNT directory and then stored with the keystrokes ['] TLST [STO] .

PROGRAMMING

```
<< →LIST >>
```

- To use this program, you should have the n data values entered in levels 2 through n+1 of the stack and the number of data values entered in level 1 of the stack.

The following program should be entered in the PCNT directory and then stored with the keystrokes ['] SORT [STO].

PROGRAMMING

```
<< IF DUP SIZE 1 > THEN LIST→ DUP 1 + ROLL 1
   →LIST LIST→ DROP → X
   << { } { } ROT 1 SWAP 1 - START ROT DUP 1 →LIST
      SWAP X IF < THEN ROT + SWAP ELSE + END
      NEXT SORT SWAP SORT X + SWAP + >>
   END DUP 'DATA' STO >>
```

- Program input is the list containing the data that has been generated by use of the program TLST.

The following program, PROMPT[1], is used by several of the programs to follow. It should be entered in the PCNT directory and then stored with the keystrokes ['] PMPT [STO].

PROGRAMMING

```
<< " ENTER " SWAP + CLLCD 1 DISP HALT >>
```

---

[1]Wickes, William C., *HP-28 Insights: Principles and Programming of the HP-28C/S*, Larken Publications, Corvallis, Oregon, 1988.

The two programs listed below are both called PCTL. The first one takes data input from the stack. The second one prompts for the value of the percentile from within the program. You should enter one of these two in your PCNT directory.

The following program should be stored with the keystrokes ['] PCTL [STO].

PROGRAMMING

```
<< SWAP DUP SIZE ROT * .01 * DUP
IP SWAP FP IF Ø > THEN 1 + GET
ELSE DUP2 1 + GET 3 ROLLD GET +
2 / END >>
```

- Program input is the list containing the data that has been generated by use of the program SORT in level 2 of the stack and the percentile you wish to find in level 1 of the stack. Recall that this list is stored under the name DATA and may be recalled to the stack by pressing [DATA]M.

OR

The following program should be stored with the keystrokes ['] PCTL [STO].

PROGRAMMING

```
<< DUP SIZE " P " PMPT * .01 * DUP
IP SWAP FP IF Ø > THEN 1 + GET
ELSE DUP2 1 + GET 3 ROLLD GET +
2 / END >>
```

- Program input is the list containing the data that has been generated by use of the program SORT in level 1 of the stack. Recall that this list is stored under the name DATA and may be recalled to the stack by pressing [DATA]M.

When the program pauses to ask you to "ENTER P", enter the desired percentile and press ■ [CONT] to continue the program.

A NOTE TO HP-28S USERS

- The box plot program and its associated subroutines for HP-28S users have not been included in this chapter due to the complexity of the programs and the lack of a built-in LINE command. It is this author's opinion that the keystroking and used memory necessary to have such programs in your calculator is not worth the effort. HP-28S users should use program PCTL to find the percentiles necessary to construct the box plot and then draw it by hand.

  HP-28S users should try the exercises at the end of this section as they apply to general concepts and the programs SORT and PCTL.

❐ SORT AND PERCENTILE PROGRAMS FOR THE HP48

Because the programs to sort data and find percentiles will also be used to construct box plots, it is best to create a directory containing these programs. Enter your statistics directory by pressing [ISTAT]M and press [MEMORY] ['] BXPD [CRDIR]M to create a subdirectory called BXPD. To see that this directory has been included in your user menu, press [VAR] and look at the menu. Press [BXPD]M to enter the BXPD directory before entering the remaining programs in this section.

The following program should be entered in the BXPD directory and then stored with the keystrokes ['] TLST [STO] .

PROGRAMMING

```
<< "n ?" "" INPUT OBJ→ →LIST >>
```

- Program input is the data that has been entered in levels 1 through n of the stack. Input the number of data values, n, when prompted.

The following program should be entered in the BXPD directory and then stored with the keystrokes ['] SORT [STO] .

PROGRAMMING

```
<< IF DUP SIZE 1 > THEN LIST→ DUP 1 + ROLL 1
   →LIST LIST→ DROP → X
   << { } { } ROT 1 SWAP 1 - START ROT DUP 1 →LIST
      SWAP X IF < THEN ROT + SWAP ELSE + END
      NEXT SORT SWAP SORT X + SWAP + >>
   END DUP 'DATA' STO >>
```

- Program input is the list containing the data that has been generated by use of the program TLST.
- HP48 users will find no menu key for LIST→. You must use the combination of keys [α] [α] L I S T [→] with no intervening spaces.

Program SORT may take a while to run if the data set is large and requires a lot of sorting. An alternate program for sorting the data is found on page 562 of Volume II of the HP48 *Owner's Manual*.

The following program should be entered in the BXPD directory and then stored with the keystrokes ['] P C T L [STO].

PROGRAMMING

```
<< DUP DUP SIZE "p ?" "" INPUT OBJ→
* .01 * DUP IP SWAP FP IF Ø > THEN
1 + GET ELSE DUP2 1 + GET 3 ROLLD
GET + 2 / END SWAP >>
```

- Program input is the list containing the data that has been generated by use of the program SORT. Recall that this list is stored under the name DATA.

If you wish only to find percentiles and not draw the box plot, programs TLST, SORT and PCTL may be used independently of the box plot program.

BOX PLOTS ON THE HP48

Box plots, sometimes called box-and-whisker plots, provide a useful graphical technique for describing data. These graphs use information regarding how the measurements are spread over the interval from the smallest value to the largest value in a set of data.

THE BOX-AND-WHISKER PLOT

The box plot is a graphical display that describes not only the behavior of the measurements in the middle of the distribution but also their behavior at the ends or tails of the distribution. A box plot graphs a five-number summary of the data: lowest value, lower quartile, median, upper quartile, highest value. Two sets of limits normally placed on the box plot are the inner and outer fences. Recall that the interquartile range, iqr, is the difference in the upper and lower quartiles. Inner fences are located a distance of 1.5 iqr below the lower quartile and 1.5 iqr above the upper quartile. Outer fences are located a distance of 3 iqr above the lower quartile and 3 iqr above the upper quartile. The ends of the box are located at the 25th and 75th percentiles with the median shown as a vertical line through the box. Horizontal lines called whiskers are drawn from the ends of the box to the smallest and largest data values inside the inner fences.

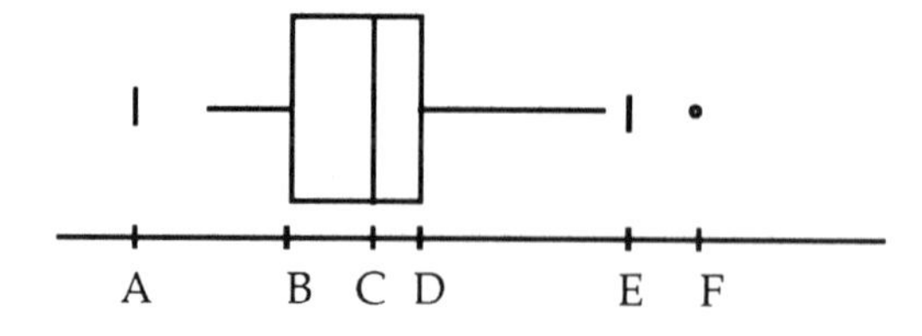

A = lower inner fence
B = 25th percentile
C = median
D = 75th percentile
E = upper inner fence
F = outlier

If the line at the median is at or near the center of the box, this is an indication of symmetry of the data. If one whisker is clearly longer than the other, the data is very likely skewed in the direction of the longer whisker. If you are in doubt as to the length of the whiskers or if you are dealing with small data sets, it is a good idea to compare the mean and median for the data to determine the skewness.

Notice that the box plot gives a graph of summary data since it uses only five statistics and does not show the individual data values as does a stem and leaf plot. We cannot identify the shape of a distribution as well as we can with stem-and-leaf plots or histograms, but we can easily look at relative positions of different sets of data and compare them quite easily. Box plots are extremely useful for very large sets of data.

Program Box plot uses the percentiles calculated by program Percentile and draws a statistical box plot of the data. The calculator range is set at the outer fences as long as no data values are outside those fences. If there are extreme data values (outliers) in either direction, the x-range uses the smallest (or highest) data value for the respective Xmin or Xmax value in the range settings. The program also displays inner fences on the graph. Program BXPT pauses after the initial box plot is drawn, and pauses again after any outliers that were found are displayed on the graph.

- Box plots may be drawn vertically as well as horizontally. Program BXPL is designed to draw horizontal box plots because of the dimensions of the calculator screen.

The remainder of the programs in this section are used by the program BXPL. Programs LBOX and RBOX set the minimum and maximum values of the x-range for drawing the box plot. Program LBL is used to display the values for the lower and upper inner fences during program execution.

The following program should be entered in the BXPD directory and then stored with the keystrokes ' LBOX STO .

PROGRAMMING

```
<< 6 PICK 2 PICK 9 PICK → b i a
   << b i - IF a > THEN a 1 - Ø
      R→C PMIN ELSE b i - Ø R→C
      PMIN END >>
>>
```

- Input is from program BXPT.

The following program should be entered in the BXPD directory and then stored with the keystrokes ' RBOX STO .

PROGRAMMING

```
<< 4 PICK 2 PICK 5 PICK → d i e
   << d i - IF e > THEN e 1 + 1.5
   R→C PMAX ELSE d i + 1.5 R→C
   PMAX END >>
>>
```

- Input is from program BXPT.

The following program should be entered in the BXPD directory and then stored with the keystrokes ' LBL STO .

PROGRAMMING

```
<< " = " + OVER →STR + CLLCD 1
DISP Ø WAIT DROP >>
```

- Input is from the program BXPT. ENTER should be pressed to resume BXPT whenever a message appears on the screen as a result of this program.

❐ BOX PLOT PROGRAM FOR THE HP48

The following program should be entered in the BXPD directory and then stored with the keystrokes ['] BXPT [STO].

PROGRAMMING

```
<< DUP 1 GET SWAP "LET p BE 25 , THEN 50 , THEN 75
   TO FIND Q1 , MED , Q3 .  PRESS ENTER TO CONTINUE . "
   CLLCD 1  DISP  Ø  WAIT  DROP  PCTL  PCTL  PCTL  CLLCD
   DUP  DUP  SIZE  GET  SWAP  6  ROLLD  2  PICK  5  PICK  -
   DUP  1.5  *  SWAP  3  *  ERASE  LBOX  RBOX  6  PICK  .375
   R→C  5  PICK  1.125  R→C  BOX  5  PICK  DUP  .375  R→C
   SWAP  1.125  R→C  LINE  DROP  5  PICK  2  PICK  -  'L'
   STO  7  PICK  Ø  → c
   << DO  'c'  INCR  GETI  'K'  STO  DROP  UNTIL  K  L  >
      END >>
   DUP  SIZE  5  PICK  4  PICK  +  'U'  STO  1  +  → c
   << DO  'c'  DECR  GETI  'M'  STO  DROP  UNTIL  M  U  <
      END >>
   6  PICK  .75  R→C  K  .75  R→C  LINE  4  PICK  .75  R→C  M
   .75  R→C  LINE  L  .72  R→C  L  .78  R→C  LINE  U  .72
   R→C  U  .78  R→C  LINE  { }  PVIEW  OTLR  { }  PVIEW  PICT
   RCL  'P'  STO  DROP  DROP  5  →LIST  DUP  "5  #  SUMMARY
   = "  CLLCD  1  DISP  2  DISP  Ø  WAIT  DROP  DATA  DROP
   L  " LIF "  LBL  U  " UIF "  LBL  { M  U  K  L }  PURGE
>>
```

- Input is the sorted list of data in level 1 of the stack.

❐ OUTLIERS

Outliers may represent faulty measurements in recording or observation or may be valid measurements which, for one reason or another, differ markedly from the others in the set. Data values that fall between the inner and outer fences are called *mild outliers*. Data values falling outside the outer fences are called *extreme outliers*. Program OTLR tests for mild and extreme outliers and indicates the outlier(s) by showing the point(s) on the box plot graph. Program OTLR is used as a subroutine of the box plot program and should not be used as an individual program for it needs information calculated in the box plot program.

The following program should be entered in the BXPD directory and then stored with the keystrokes ['] OTLR [STO]. Input is from program BXPT.

PROGRAMMING

```
<< DUP SIZE 1 SWAP FOR i i GETI 'OL'
STO DROP IF OL L < OL U > XOR
THEN OL .75 R→C PIXON END 'OL'
PURGE NEXT >>
```

When you first see the box plot on the display screen, press [ON] to resume execution of the program. Program OTLR is then called automatically by the box plot program and you will see again see the graph of the box plot. [ON] resumes execution of the program to give the five number summary. Press [ENTER] until the program finishes giving you information. You may then press [◁] to recall the graph to the display screen. Look closely at the graph to see if any outliers have been indicated. If so, you can use the left and right arrow keys to move the crosshair cursor to the position of the outlier. Press [+] to obtain a pixel coordinate estimate of the outlier (the x coordinate). Refer to the data for the exact value of that point.

After the first time you use the programs in the BXPD directory, you may wish to reorder the VAR menu for easier access in the future. Press [{ }] [TLST]M [SORT]M [PCTL]M [BXPT]M [DATA]M [LBOX]M [RBOX]M [OTLR]M [LBL]M [P]M [PPAR]M [ENTER] [MEMORY] [ORDER]M and your BXPD directory will placed in the order in which you entered the programs in the above list.

EXERCISES

1. The number of times each letter of the English alphabet occurred was counted on one randomly chosen page of a statistics text. The table below lists each of the letters and the percentage that each letter occurred.

A	7.9	H	5.6	O	6.5	V	0.8
B	1.8	I	6.6	P	2.1	W	1.6
C	2.6	J	0.1	Q	0.09	X	0.15
D	3.3	K	0.4	R	6.5	Y	2.4
E	13.7	L	3.7	S	7.3	Z	0.06
F	2.7	M	2.6	T	10.0		
G	2.0	N	8.0	U	1.5		

   a) Use program SORT to enter this data and sort it in increasing order.
   b) While you are waiting for the calculator to finish, answer the following questions:
      i) What is the most-used letter? the least-used letter?
      ii) What percentage of the letters used are vowels?
      iii) Why do you think the letters T, R, S, L, N and E are the ones most often chosen on the *Wheel of Fortune* game show?
   c) Find the quartiles and the median percentage.
   d) Construct a box plot of the percentages.
   e) Are most of the letters used rarely or frequently? What conclusion can you draw from the length of the whiskers on your box plot?
   f) Identify any outliers.

2. Choose one page at random from your statistics text and one page at random from your English text. (If you choose a page from either with more than one illustration, choose another at random.) Count the occurrence of the letters on each page and calculate the percentage of occurrence of each of those letters. Answer the same questions that are in problem 1 above for each set of data. Compare the box plots.

3. The following are the numbers of private aircraft which landed at a large metropolitan airport on fifteen consecutive days:.

   85 74 67 87 71 89 82 125 73 84 77 82 70 90 38

   Construct a box plot for this data and interpret the graph. Give the five number summary. Find and discuss the possible reasons for any outliers that may be present.

4. Construct a box plot for the cigarette tax data given on page 205 of this manual. Are there any outliers?

ANSWERS

1. b) i) The most used letter is E and the least used letter is Z.
   ii) The letters A, E, I, O and U are used 36.2% of the time.
   iii) These letters are normally chosen since they have the highest frequency of occurrence in most phrases.

   c) There are n = 26 data values. The 25th percentile, Q_1, is 1.5. The 75th percentile, Q_3, is 6.5. The median, the 50th percentile, is 2.6.

   d) Note that the interquartile range = 5, lower inner fence = -6 and upper inner fence = 14. The box plot is

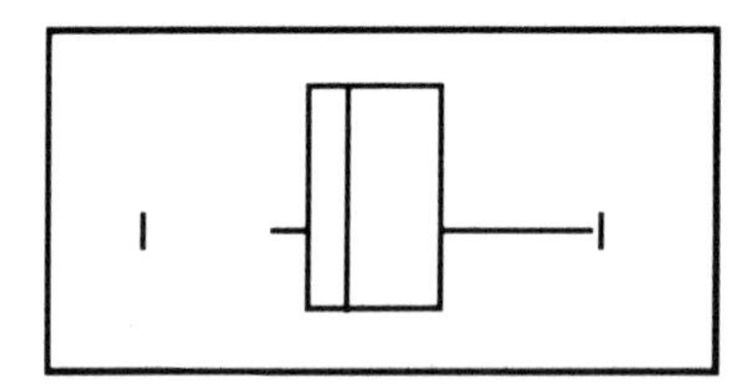

   e) Since half of the letters are used about 2.6% of the time or less, most are rarely used. Since the right whisker is much longer than the left, we conclude that the range of the data from Q_1 to the lowest data value, 0.06 ,is much less than the range of the data from Q_3 to the highest data value, 13.7. Note that the middle half of the data is spread from 1.5 to 6.5. The data is positively skewed.

   f) There are no outliers in this set of data.

2. In this problem, you should use the calculator to aid you in the construction of the box plots for the two sets of data and then reconstruct them by hand using the same scale on a single set of axes. Look at the lengths of the boxes, the location of the medians and the lengths of the whiskers to determine whether of not the variabilities in the two sets of data are approximately the same or different. Check to see if the pattern of outliers is the same in both data sets if outliers are detected. Note that it does not matter whether or not the number of data values is the same for both groups of data.

3. There are n = 15 data values. The smallest data value is 38, the 25th percentile, Q_1, is 71, the median, the 50th percentile, is 82, the 75th percentile, Q_3, is 87 and the largest data value is 125. Note that the interquartile range equals 16, the lower inner fence is 47 and upper inner fence is 111. The box plot is

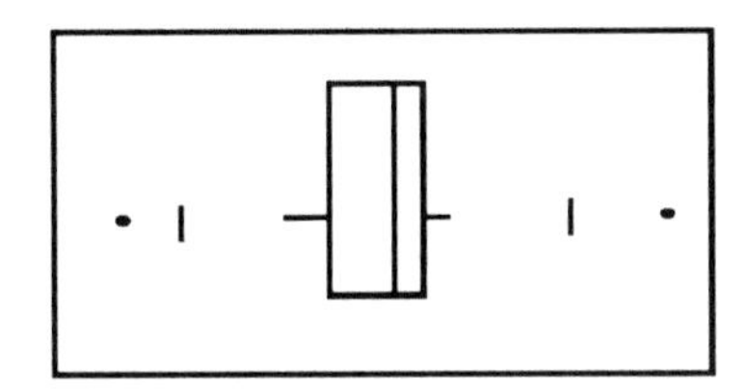

There is an indication that the data are negatively skewed since the left whisker is longer than the right whisker. You should check the relationship between the mean and the median to verify this, however. Note that approximately one-half of all the observations lie within the box. Moreover, about one-fourth of the observations lie between the median and one side of the box and approximately one-fourth lie between the median and the other side of the box.

You should also consider the practical interpretations of this data and with your group discuss what these numbers would mean to airport controllers, planning for additional runways, etc. Discuss these interpretations and the reasons for the outliers with other students.

4. The box plot is shown below. Notice that there are no outliers indicated.

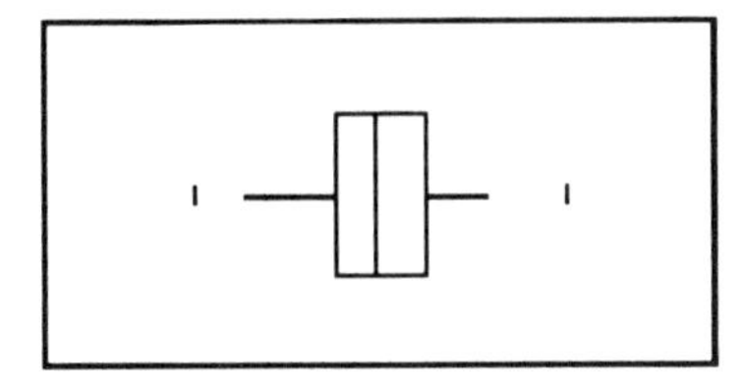

MULTI-VARIABLE DATA

There may be times that you wish to enter and investigate data consisting of more than one variable. The matrix capabilities of your Hewlett-Packard calculator make this task relatively easy.

ENTERING AND EDITING MULTI-VARIABLE DATA ON THE HP-28S

Before entering new data, be certain that you have cleared the current statistical matrix with CLΣ M. Multi-variable data is entered into the calculator such that each column (vertical) represents a different variable. The number of rows that are entered equals the number of data points. To enter data consisting of more than one variable, prefix each row of data with [, separate the data values with commas, and press Σ+ M after each entry. Consider the following example:

Professor Ogle teaches four statistics classes. The number of absences for each of the classes during the first five class meetings in April are as follows:

	April 1	April 3	April 5	April 8	April 10
8:00 a.m. class	2	1	3	1	0
10:00 a.m. class	0	1	2	1	1
1:00 p.m. class	1	0	3	2	0
3:30 p.m. class	3	2	5	2	0

To enter the first row of data (the 8 a.m. class), press [2 , 1 , 3 , 1 , 0 , Σ+ M. The second row of data (the 10 a.m. class) is entered in exactly the same manner. Continue until all rows are entered. To view the matrix, press RCLΣ M and if necessary, ■ VIEW↓. Refer to page 133 of your *Owner's Manual* for instructions on editing the data.

The built-in statistical functions in the STAT menu are available for this data. Summary statistics such as the mean and standard deviation will be given for all columns when you press the proper keys in the STAT menu. These will be given in the form of a vector with the individual statistics in the order of the columns. For example, suppose you wished to find the mean number of absences on April 5. Press [MEAN]^{M} and you see a 5 element vector of means. Since April 5 is listed as the third column, look in the third position of the means vector to find 3.25. How would you find the mean number of absences for the 10 a.m. class? You now need the mean of the first row, not a column. With your matrix of original data in level 1 of the stack, press ■ [ARRAY] [TRN]^{M} [STOΣ]^{M}. You will see that the rows and columns have been switched and that you may now obtain the mean number of absences for the 8 a.m. class by pressing [MEAN]^{M} and choosing the first element, 1.4.

❒ ENTERING AND EDITING MULTI-VARIABLE DATA ON THE HP48

Before entering new data, be certain that you have cleared the current statistical matrix with [CLΣ]^{M}. Multi-variable data is entered into the calculator such that each column (vertical) represents a different variable. The number of rows that are entered equals the number of data points. To enter data consisting of more than one variable, enter *only the first row* of data with brackets ([↰] [×]), separate the data values with a space, and press [Σ+]^{M} after each entry.

Refer the example on the previous page about the four statistics classes. To enter the first row of data (the 8 a.m. class), press [↰] [×] 2 [SP] 1 [SP] 3 [SP] 1 [SP] 0 [SP] [Σ+]^{M}. The second row of data (the 10 a.m. class) is entered with 0 [SP] 1 [SP] 2 [SP] 1 [SP] 1 [SP] [Σ+]^{M}. Enter the remaining rows as you did the second row. To view the matrix, press [EDITΣ]^{M}. Refer to page 370 of your *Owner's Manual* for instructions on editing the data.

The built-in statistical functions in the STAT menu are available for this data. Summary statistics such as the mean and standard deviation will be given for all columns when you press the proper keys in the STAT menu. These will be given in the form of a vector with the individual statistics in the order of the columns. For example, suppose you wished to find the mean number of absences on April 5. Press MEAN M and you see a 5 element vector of means. Since April 5 is listed as the third column, look in the third position of the means vector to find 3.25. How would you find the mean number of absences for the 10 a.m. class? You now need the mean of the first row, not a column. With your matrix of original data in level 1 of the stack, press MATH MATR M TRN M STOΣ M. You will see that the rows and columns have been switched and that you may now obtain the mean number of absences for the 8 a.m. class by pressing MEAN M and choosing the first element, 1.4.

WORKING WITH THE STATISTICAL DATA MATRIX

In the event it is necessary to extract a particular column from a multi-variable statistical data matrix, the following program can be used. The following program should be entered in one of your directories and then stored with the keystrokes ' GETC STO. The program is given in two slightly different versions for the two different HP calculators. GETC stores the data you originally entered under the name ORIG and stores the extracted single column as the current statistical matrix.

PROGRAMMING *for HP-28S Users*

```
<< SWAP DUP DUP 'ORIG' STO TRN STOΣ SWAP
   1 + NΣ DUP2 IF ≤ THEN START Σ- DROP
   NEXT ELSE DROP2 END Σ- ARRY→ LIST→ 2
   →LIST →ARRY DUP STOΣ >>
```

- Input from level 2 of the stack is the current statistical matrix composed of at least two columns. Input from level 1 of the stack is the number of the column you wish to extract.

You may edit this single column of data and/or work with it as you would if you had entered the single-variable data. If you wish to find percentiles for data in this matrix, you must extract the column you wish to use, convert it to a list ({ }) with the keystrokes [ARRY→]M [LIST→]M [DROP] [DROP] [→LIST]M before running program SORT.

PROGRAMMING *for HP48 Users*

```
<< SWAP DUP DUP 'ORIG' STO TRN STOΣ SWAP
   1 + NΣ DUP2 IF ≤ THEN START Σ- DROP
   NEXT ELSE DROP2 END Σ- OBJ→ OBJ→ 2
   →LIST →ARRY DUP STOΣ >>
```

- Input from level 2 of the stack is the current statistical matrix composed of at least two columns. Input from level 1 of the stack is the number of the column you wish to extract.

You may edit this single column of data and/or work with it as you would if you had entered the single-variable data. If you wish to find percentiles and/or construct box plots for data in this matrix, you must extract the column you wish to use, convert it to a list ({ }) with the keystrokes [OBJ→]M [OBJ→]M [DROP] [DROP] [→LIST]M before running program SORT.

For most statistical applications, however, you may use the multi-variable ΣDAT matrix *without* having to extract the particular column containing the data you wish to use.

HP-28S USERS

You may specify certain columns for use with statistical functions. Press NEXT until you see COLΣ[M]. Enter on the stack the number for the x column followed by the number for the y column and separate the two values with a comma. Press COLΣ[M]. When you use the histogram programs, histograms are drawn for the data in the column that is specified as the x column. To draw a scatter diagram, choose the x and y columns you wish to use and press DRWΣ[M] (in the PLOT menu).

HP48 USERS

You may specify certain columns for use with statistical functions. Press NXT until you see the menu containing XCOL[M] and YCOL[M]. There will be a message at the top of the display screen telling you which columns are currently chosen. To change the specifications, enter the column number on the stack and press the appropriate key. When you use BARPL[M] or HISTP[M], histograms are drawn for the data in the column that is specified as the x column. To draw a scatter diagram, choose the x and y columns you wish to use and press SCATR[M] (in the STAT menu).

ALL HP USERS

Vectors are one dimensional arrays of numbers which appear on your calculator screen enclosed in brackets. Vectors of the same dimension may be added, subtracted, and multiplied by real numbers. To extract a particular column of your ΣDAT matrix in vector form, the following portion of program GETC may be used. Input from the stack is the same as for program GETC.

```
<< SWAP DUP DUP 'ORIG' STO TRN STOΣ SWAP
   1 + NΣ DUP2 IF ≤ THEN START Σ- DROP
   NEXT ELSE DROP2 END Σ- >>
```

❑ PROJECT: UNITED STATES PRESIDENTS

The following[1] is a list of Presidents of the United States, age at which each was first inaugurated, the age at death, and the number of children[2] :

	President	Age Inaugurated	Age Died	Sons/Daughters	
1.	Washington	57	67	none	
2.	J. Adams	61	90	3 / 2	
3.	Jefferson	57	83	1 / 5	
4.	Madison	57	85	none	
5.	Monroe	58	73	0/ 2	
6.	J. Q. Adams	57	80	3 / 1	
7.	Jackson	61	78	none	
8.	Van Buren	54	79	4 / 0	
9.	W. H. Harrison	68	68	6 / 4	
10.	Tyler	51	71	3 / 5	(5 / 2)
11.	Polk	49	53	none	
12.	Taylor	64	65	1 / 5	
13.	Fillmore	50	74	1 / 1	(0 / 0)*
14.	Pierce	48	64	3 / 0	
15.	Buchanan	65	77		
16.	Lincoln	52	56	4 / 0	
17.	A. Johnson	56	66	3 / 2	
18.	Grant	46	63	3 / 1	
19.	Hayes	54	70	7 / 1	
20.	Garfield	49	49	4 / 1	
21.	Arthur	50	56	2 / 1	
22.	Cleveland	47	71	2 / 3	
23.	B. Harrison	55	67	1 / 1	(0 / 1)
24.	McKinley	54	58	0 / 2	
25.	T. Roosevelt	42	60	0 / 1	(4 / 1)
26.	Taft	51	72	2 / 1	
27.	Wilson	56	67	0 / 3	(0 / 0)
28.	Harding	55	57	0 / 0	
29.	Coolidge	51	60	2 / 0	
30.	Hoover	54	90	2 / 0	
31.	F. Roosevelt	51	63	4 / 1	
32.	Truman	60	88	0 / 1	

[1] *World Almanac and Book of Facts*, Pharos Books, New York, 1991.

[2] Deceased infants are not included.

*Married twice.

33.	Eisenhower	62	78	1 / 0	
34.	Kennedy	43	46	1 / 1	
35.	L. Johnson	55	64	0 / 2	
36.	Nixon	56		0 / 2	
37.	Ford	61		3 / 1	
38.	Carter	52		3 / 1	
39.	Reagan	69		1 / 1	(1 / 1)
40.	Bush	64		4 / 2	

The above statistical data may be entered in a 40 by 4 ΣDAT matrix for further study. Clear the current statistical matrix and enter the data as described on the previous pages (use zeros to indicate no sons and daughters for George Washington). Continue entering data through the 35th President, Lyndon Johnson. For Nixon through Bush, enter 0's for the age of death. (We will change this later.) For those presidents who have been married more than once, enter the total number of sons in the third column and the total number of daughters in the fourth column. Store this matrix under the name PRES.

EXERCISES

1. Extract columns 1 and 2 of the ΣDAT matrix PRES and delete the last 5 elements of column 2. (Recall that these were the 0's we filled in as the age of death in order to enter the original multi-variable data matrix.)

   a) What is the mean age at inauguration for U. S. Presidents? the mean age of death? Which of these two ages has the greatest variability? Why?

   b) Construct box plots for the age at inauguration and the age of death. Write a brief paragraph describing any similarities and/or differences in the box plots. Include a sketch, using one set of axes, of both the box plots.

   c) Construct histograms of 8 classes of equal width for the inauguration age data and the age of death data. Does the skewness appear the be the same as determined by the box plots?

   d) Choose the age of inauguration and the age of death as the respective x and y columns for your matrix. Construct a scatter diagram. Can you, from the scatter diagram, make any predictions about the age of death of the living U. S. Presidents?

2. Extract columns 3 and 4 of the ΣDAT matrix PRES.

   a) What is the mean number of sons for U. S. Presidents? the mean number of daughters? Which of these two has the greatest variability?

   b) Construct box plots for the number of sons and the number of daughters. Write a brief paragraph describing any similarities and/or differences in the box plots. Include a sketch, using one set of axes, of both the box plots.

   c) Construct histograms of 8 classes of equal width for the number of sons data and the number of daughters data. Does the skewness appear the be the same as determined by the box plots?

   d) Choose the number of sons and the number of daughters as the respective x and y columns for your matrix. Construct a scatter diagram. Is any pattern visible in the scatter diagram?

3. If you are working on this as a group project, discuss with students in your group other questions that could be asked about this presidential data. Choose what you consider the most interesting question and try to determine how to use the calculator to obtain an answer.

PROBABILITY DISTRIBUTIONS WITH THE HP-28S

Probabilities for the binomial, Poisson and other probability distributions are readily obtained from the algebraic equation of the distribution as entered into a program on your calculator which stores the probabilities in the statistical matrix ΣDAT. This matrix may be renamed and stored for future reference if desired.

Unfortunately, due to the lack of a built-in program for drawing histograms of relative frequencies on the HP-28S, the graphs of discrete probability distributions are not easily generated. (The histogram program given previously cannot be used because the bins contain frequencies, not probabilities.) HP users should refer to Chapter 4 of this manual for general information regarding discrete probability distributions and their graphs.

Because there are several programs used in this section, it would be helpful to create a directory to contain them. Enter your statistics directory by pressing [ISTAT]M and press [MEMORY] ['] D S C R [CRDIR]M to create a subdirectory called DSCR. To see that this directory has been included in your user menu, press [USER] and look at the menu. Press [DSCR]M to enter the discrete graphs directory before entering the remaining programs in this section. You should include program [QUIT] as the first program entered in this new directory.

BINOMIAL PROBABILITY DISTRIBUTION

The program below will give you a convenient way of calculating all the binomial probabilities p(x) for x = 0, 1, 2, . . . , n. The program stores the probabilities that are calculated in the matrix ΣDAT. After executing the program, press ΣDAT and you will see the matrix of probabilities which contains P(x = i) in row i+1 of the matrix.

If you wish to access the built-in statistical functions in the STAT menu, press [USER] [ΣDAT]M to place the matrix of probabilities on the stack and key in

■ [STAT] [CLΣ]^M [STOΣ]^M to make ΣDAT the current statistical matrix. The reason that you must clear the matrix ΣDAT in the STAT mode is that it will probably be a different matrix than the one you created in the DSCR directory even though it has the same name.

- Notice that the mean of the *probabilities* can be found by pressing [MEAN]^M and the standard deviation is given by pressing [SDEV]^M. These, however, are *not* the mean and standard deviation of the probability distribution for the values of x have not been considered. You may modify the programs in this section to store the x-values as well as the probabilities in a two-dimensional matrix ΣDAT or use the "short-cut" formulas in your text to obtain μ and σ. These formulas could be included as output of the probability distribution programs in this section as was done for the Sharp EL-5200 in Chapter 4 of this manual.

The following program, binomial probability distribution, should be entered in the DSCR directory and then stored with the keystrokes ['] B I P D [STO] .

PROGRAMMING

```
<<  CLEAR  ∅  DUP  →  n  p
    <<  {  STO  n  p  }  MENU  HALT  CLΣ  ∅  n  FOR  k  n  k
        COMB  ' C '  STO  ' C  *  p  ^  k  *  (1  -  p )  ^  ( n  -  k )  '
    EVAL  Σ+  NEXT
    >>  ΣDAT  ' C '  PURGE
>>
```

- This program requires no user input *before* pressing [BIPD]^M.

- When this program is executed by pressing BIPD^M, you will see a menu containing the letters N and P. It is important to note that these variables must be entered as local variables (that is, using lowercase letters) in the program above. They just happen to appear on the menu as capital letters.

- To run this program, enter the value of n, the number of trials and press N^M to store this value as n. Enter the value of p, the probability of success on each trial, and press P^M to store this value as p. Press ■ CONT to continue the program and generate the matrix of probabilities.

- To view the entire matrix, press ' ΣDAT^M ■ VISIT . You may use the cursor key ▽ to scroll through the contents of the matrix. Press ON to return to the stack environment.

❒ POISSON PROBABILITY DISTRIBUTION

The program below, as in the case of the binomial distribution discussed in the previous section, gives you a convenient way of calculating the Poisson probabilities p(x) for x = 0, 1, 2, . . . , 10λ.

The following program, Poisson distribution probabilities, should be entered in the DSCR directory and then stored with the keystrokes ' PODP STO .

PROGRAMMING

```
<< CLEAR  ∅  DUP  →  ℓ  k
   << {  STO  ℓ }  MENU  HALT  CLΣ  ∅  10  ℓ  *  FOR
      k  k  FACT  ' F '  STO  ' ℓ ^ k * EXP ( CHS  ℓ )  /  F '
      EVAL  Σ+  NEXT  >>
   ΣDAT  >>
```

- The character ℓ is obtained by pressing [LC] L.

- This program requires no user input *before* pressing [PODP]M.

- When this program is executed with [PODP]M, you will see a menu containing the symbol L.

- To run this program, enter the value of ℓ, the expected number of successes, and press [L]M to store this value as ℓ. Note that $\ell = \lambda$ which is represented in the menu as L. Press ■ [CONT] to continue the program and generate the probabilities.

EXERCISES

1. Generate the binomial distribution probabilities for $n = 10$ using the program BIDP. Compare these values with those from the table in your text. Enter the STAT mode, make your ΣDAT matrix the current statistical matrix, choose and verify that the sum of the probabilities is 1.

2. Use the program PODP to generate probabilities for the Poisson distributions with $\lambda = 2.63$ and $\lambda = 8$. In each case, enter the STAT menu, make your ΣDAT matrix the current statistical matrix and press [TOT]M which will give you the sum of the probabilities. Do they sum to 1 as they should? Why or why not?

3. Consult your text for the defining formula for hypergeometric probabilities and/or geometric probabilities. (You can also refer to page 94 of this manual for a discussion of hypergeometric probabilities.) Use the methods presented for the binomial and Poisson probabilities to construct a program to generate the probabilities for the hypergeometric or geometric distributions. You may wish to read about local (lowercase letters) and global (uppercase letters) variables in your *Owner's Manual* before you attempt writing your program.

PROBABILITY DISTRIBUTIONS WITH THE HP48

Graphs of the binomial, Poisson and hypergeometric probability distributions are readily obtained by programming the algebraic equation of the distribution. The probabilities for the distribution are generated by the program and stored in the statistical matrix ΣDAT. This matrix may be renamed and stored for future reference if desired. HP users should refer to Chapter 4 of this manual for information regarding discrete probability distributions and their graphs.

Because there are several programs used in this section, it would be helpful to create a directory to contain them. Enter your statistics directory by pressing [ISTAT]M and press [MEMORY] ['] DSCR [CRDIR]M to create a subdirectory called DSCR. To see that this directory has been included in your user menu, press [VAR] and look at the menu. (HP-28S users should press [USER] to see the user menu and should include program [QUIT] as the first program entered in this new directory.) Press [DSCR]M to enter the discrete graphs directory before entering the remaining programs in this section.

BINOMIAL PROBABILITY DISTRIBUTION

The defining equation of the binomial probability distribution can be entered as indicated by the following keystrokes or can be entered as it appears in your textbook into the EquationWriter and placed in the program below. Access the Equation Writer with [⬑] [EQUATION] and enter the binomial distribution formula (page 64 of this manual) to obtain

$$\text{COMB}(n,k)\, p^{k}\, (1-p)^{n-k}$$

Press ENTER and the equation is copied to the stack, ready for insertion into the program. Refer to your *Owner's Manual* if you are not familiar with the EquationWriter application.

Equations for probability distributions in the programs in the section may be obtained in this manner. Programming techniques for finding specific probabilities, not entire probability distributions, are given in Chapter 12 of this manual.

The program below will graph the binomial distribution probability histogram for the number of successes, x, between 0 and n. Notice that it also gives you a convenient way of calculating all the binomial probabilities p(x) for x = 0, 1, 2, . . . , n. The program stores the probabilities that are calculated in the matrix ΣDAT. After the graph is drawn, press ΣDAT and you will see the matrix of probabilities which contains P(x = i) in row i+1 of the matrix.

If you wish to access the built-in statistical functions in the STAT menu for data that was previously entered and is not in the current statistical matrix, press VAR ΣDAT M (or whatever name you have called the data) to place the matrix of probabilities on the stack and key in ◁ STAT STOΣ M to make your ΣDAT active. If you prefer, this matrix could also be called to the stack and stored as the current statistical matrix using the CAT M command in the STAT menu.

- Notice that the mean of the *probabilities* can be found by pressing MEAN M and the standard deviation is given by pressing SDEV M. These, however, are *not* the mean and standard deviation of the probability distribution because the values of x have not been considered. You may modify the programs in this section to store the x-values as well as the probabilities in a two-dimensional matrix ΣDAT or use the "short-cut" formulas in your text to obtain μ and σ.

The following program, binomial distribution graph, should be entered in the DSCR directory and then stored with the keystrokes ['] BIDG [STO] .

PROGRAMMING

```
<<  Ø DUP → n p
    << { n p } MENU HALT CLΣ Ø n FOR k ' COMB ( n , k)
       * p y^x k * (1 - p ) y^x ( n - k ) ' EVAL Σ+ NEXT
    >> BARPL GRAPH Ø MENU
>>
```

- This program requires no user input *before* pressing [BIDG]^M.

- When this program is executed with [BIDG]^M, you will see a menu containing the letters N and P. It is important to note that these variables must be entered as local variables (that is, using lowercase letters) in the program above. They just happen to appear on the menu as capital letters.

- To run this program, enter the value of n, the number of trials and press [⬑] [N]^M to store this value as n. Enter the value of p, the probability of success on each trial, and press [⬑] [P]^M to store this value as p. Press [⬑] [CONT] to continue the program and draw the graph.

EXERCISES

1) Use the program BIDG to construct graphs for the binomial distributions with n = 10 and p = 0.10, n = 10 and p = 0.35, n = 10 and p = 0.60, and n = 10 and p = 0.90 . How is the shape of the graph changing as p increases?

2) Use the program BIDG to construct graphs for the binomial distributions with n = 5 and p = 0.30, with n = 15 and p = 0.30 and with n = 25 and p = .30. How is the shape of the graph changing as the value of n increases?

POISSON PROBABILITY DISTRIBUTION

The defining equation of the Poisson probability distribution (refer to page 87 of this manual) may be entered in the EquationWriter application and copied into the following program or can be keyed in entirely from the keyboard. The program below will graph the Poisson distribution probability histogram for the number of successes, x, between 0 and 10λ. Notice that the program, as in the case of the binomial distribution discussed in the previous section, gives you a convenient way of calculating the Poisson probabilities p(x) for x = 0, 1, 2, . . . , 10λ.

The following program, Poisson distribution graph, should be entered in the DSCR directory and then stored with the keystrokes ['] PODG [STO] .

PROGRAMMING

```
<< Ø DUP → λ k
   << { λ } MENU HALT CLΣ Ø ' 10 * λ '
      FOR k ' λ y^x k * EXP ( - λ ) / k ! '
      EVAL Σ+ NEXT >>
   BARPL GRAPH 2 MENU
>>
```

- The character λ is obtained by pressing [α] [↱] [NXT].
- This program requires no user input *before* pressing [PODG]M.
- When this program is executed with [PODG]M, you will see a menu containing the symbol λ.
- To run this program, enter the value of λ, the expected number of successes, and press [↰] [λ]M to store this value as λ. Press [↰] [CONT] to continue the program and draw the graph.

EXERCISES

1. Use the program PODG to construct graphs for the Poisson distributions with $\lambda = 1.75$, $\lambda = 5$ and $\lambda = 10$. How is the shape of the graph changing as λ increases?

2. Use the program PODG to construct graphs for the Poisson distributions with $\lambda = 2.63$ and $\lambda = 8$. In each case, enter the STAT menu and press TOT M which will give you the sum of the probabilities. Do they sum to 1 as they should? Why or why not?

3. Consult your text for the defining formula for hypergeometric probabilities and/ or geometric probabilities. (You can also refer to page 92 of this manual for a discussion of hypergeometric probabilities.) Use the methods presented for the binomial and Poisson probabilities to construct a program to generate the graphs of the hypergeometric or geometric distributions. You may wish to read about local (lowercase letters) and global (uppercase letters) variables in your *Owner's Manual* before you attempt writing your program.

NORMAL DISTRIBUTION OVERLAY

If you have not already done so, you soon will work with the normal distribution. The normal distribution has been studied extensively and can in many cases be used to approximate probabilities for discrete random variables. Your statistics text will very likely give you "rules-of-thumb" as to when these approximations are valid. Geometrically, whenever one considers an approximation by the normal distribution, the bell-shaped curve of the normal probability distribution should "fit" nicely over the graph of the distribution it is approximating. The following program, NDST, draws a graph of the normal distribution over the probability histogram you have generated using the program BIDG. The normal random variable should have the same mean and standard deviation as the binomial random variable for which you have drawn the probability histogram.

The following program, Normal distribution graph, should be entered in the DSCR directory and then stored with the keystrokes ['] NDST [STO].

PROGRAMMING

```
<<  Ø  DUP  →  μ  σ
    <<  { μ σ }  MENU  HALT  ' EXP ( - ( X- ( μ + .5) )
        y^x 2 / ( 2 * σ y^x 2 ) ) / ( σ * √( 2 * π) ) '
        STEQ  FUNCTION  DRAW  GRAPH  >>
    { X EQ PPAR }  PURGE 2  MENU
>>
```

- The character μ is obtained by pressing [α] [↱] [N] and the character σ is obtained by pressing [α] [↱] [S].
- This program requires no user input *before* pressing [NDST]M.
- When this program is executed with [NDST]M, you will see a menu containing the symbols μ and σ.
- To run this program, enter the value of μ, the mean of the normal distribution, with the keystrokes [↰] [μ]M and enter the value of σ, the standard deviation of the normal distribution, with the keystrokes [↰] [σ]M. Press [↰] [CONT] to continue the program and draw the graph.

If you execute the program NDST before you erase the contents of the graphics screen, the normal distribution graph will be drawn on top of the graph of the binomial distribution. You may press [◁] to retrieve the binomial graph with its normal distribution overlay at any time before another graph is drawn.

As an example, let's draw the graph of the binomial distribution for $n = 12$ and $p = .45$ using the program BIDG with the following keystrokes:

[BIDG]^M 12 [⬑] [N]^M . 45 [⬑] [P]^M [⬑] [CONT] and after viewing the graph, press [ON] [NDST]^M 12 [ENTER] . 45 [X] [⬑] [μ]^M 12 [ENTER] . 45 [X] 1 [ENTER] . 45 [−] [X] [√] [⬑] [σ]^M [⬑] [CONT]. When the graph of the binomial distribution reappears on your screen, watch the overlay of the normal distribution whose mean and variance you have specified to be the same as that of the binomial distribution. You should see

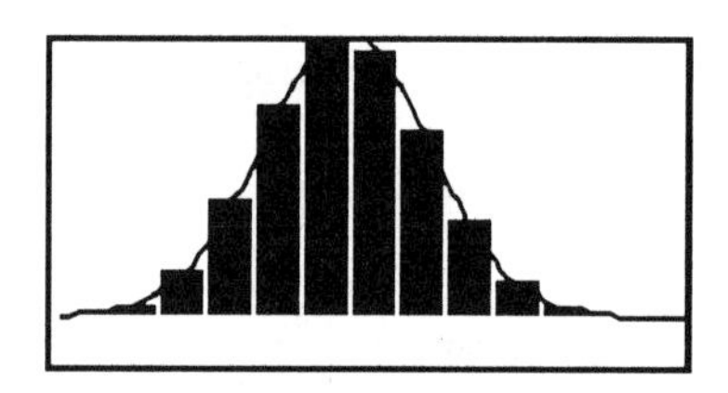

- Notice that the mean and standard deviation of the normal distribution used above were determined by the "short-cut" formulas

$$\mu = np \text{ and } \sigma = \sqrt{n\,p\,(1-p)}\,.$$

- If you look closely at the equation of the normal density function in the program NDST, you will notice that the graph has been shifted to the right 0.5 units. Whenever you are graphing both a discrete probability distribution and continuous probability distribution on the same set of axes, you must use what is called the *continuity correction*. Ask your instructor or consult your statistics text for further explanation.

Unfortunately, a problem arises when using your HP to overlay the normal distribution on discrete distribution graphs composed of many rectangles. Due to the autoscaling of the barplot function and the fact that may rectangles are being placed in so small an area, the discrete distribution graph does not always appear in the position it should. For instance, use the program BIDG to construct binomial distribution graphs for n = 18, p = .5, n = 45, p = .5, and n = 100, p = .5. Each of these graphs should appear in the center of the screen because each is symmetric about its mean which is very close to the middle of the x range. In each case, overlay the normal distribution with the appropriate mean and variance. The normal distribution, being graphed as an equation, appears in the correct position on the display screen. The normal distribution should "fit" directly over these graphs but does not. Be wary of this when you are using the normal distribution overlay of any discrete distribution graph composed of numerous rectangles. *Be certain that you look at the shape of the discrete distribution, not its position on the screen.*

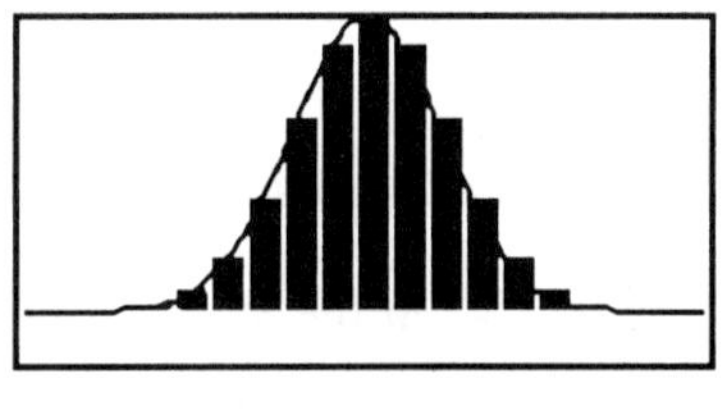

n = 18, p = .5

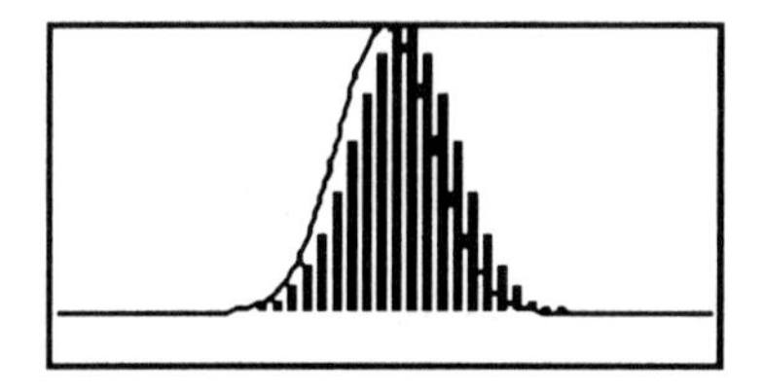

n = 45, p = .5

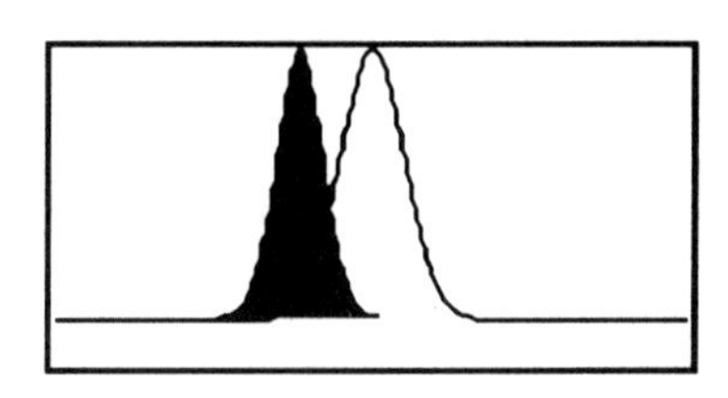

n = 100, p = .5

EXERCISES

1) Use the program BIDG to construct graphs for the binomial distributions with $n = 5$ and $p = 0.20$ and with $n = 15$ and $p = 0.56$. In each case, overlay the graph of the normal distribution with the program NDST. Which bell-shaped curve best fits (in terms of the areas under the two being nearly the same) the underlying binomial distribution?

2) Use the program PODG to construct graphs for the Poisson distributions with $\lambda = 2.63$ and $\lambda = 8$. In each case, overlay the graph of the normal distribution with the program NDST. Which bell-shaped curve best fits (in terms of the areas under the two being nearly the same) the underlying Poisson distribution?

3) a) Construct a graph of the binomial distribution for $n = 50$ and $p = 0.1$. Save the graph as B and the ΣDAT matrix containing the probabilities under a name of your choice. Determine the skewness, mean and standard deviation of the distribution.

b) Construct a graph of the Poisson distribution for $\lambda = 5$. Save the graph as P and the ΣDAT matrix containing the probabilities under a name of your choice. Determine the skewness, mean and standard deviation of the distribution.

- You can "join" two graphs that have been saved by placing one on top of the other using the following procedure: Enter B on the stack by pressing [B]M and enter P on the stack by pressing [P]M. Press [+] [SEE]M and use [◁] to recall the "joined" graph to the screen.

Does it appear that the graph of the Poisson distribution would "fit" over the graph of the binomial distribution you constructed in a)? What is the relation of the mean of the binomial to the mean of the Poisson in this problem?

c) Repeat parts a) and b) with $n = 100$.

CHAPTER 12

SIMULATION AND SOLVING ON THE HP-28S AND HP48

SIMULATION

Simulation techniques on the Hewlett-Packard calculators may be programmed with stack manipulations or in the form of algebraic expressions. The basics of simulation techniques are discussed on pages 52 through 58 of this manual and should be read for content by HP users.

Let's begin our look at simulation by going to the racetrack. Enter the following two programs on your HP 48:

Coin Race

```
<< 1 RES ERASE 0 DUP → n
   << { n } MENU HALT
   CLΣ 1 n FOR k RAND
   RDZ RAND 2 * IP RAND
   2 * IP RAND 2 * IP
   RAND 2 * IP RAND 2
   * IP + + + +
   Σ+ 0 6 XRNG
   -2 n .5 * YRNG
   HISTOGRAM DRAW
   NEXT >>
   GRAPH 2 MENU 0 RES >>
```

Die Race

```
<< 1 RES ERASE 0 DUP
   → n << { n } MENU
   HALT CLΣ 1 n
   FOR k RAND 6 *
   1 + IP Σ+ ΣDAT
   1 7 XRNG -1 n
   6 / 5 + YRNG
   HISTOGRAM DRAW
   NEXT >>
   GRAPH 2 MENU
   0 RES >>
```

As you will see when you run each of these programs, a histogram of 6 rectangles is drawn one vertical block at a time. Think of each rectangle in the histogram as the moves for one horse on the track. Thus, you are watching a race for 6 horses. Think of each block in the rectangles as a horse moving one unit. The total number of units moved by all the horses is N.

To execute each of the above programs, press the menu key in whose name you have stored the program. Load the value of N by entering the numerical value you wish on the stack and pressing [↰]. Press [↰] [CONT] to continue the program and run the 6 horse race. Gook luck on picking the winner!

EXERCISES

Refer to the information pertaining to simulating the toss of a fair coin and the roll of a fair die on pages 52 through 56 in this manual and answer the following:

1. How many coins are being tossed in the Coin Race horse race program?

2. If x is the random variable representing the number of heads obtained in the toss of these coins, what are the possible values for x?

3. How many dice are being rolled in the Die Race horse race program?

4. If y is the random variable representing the number of dots appearing on the upturned face of the die, what are the possible values for y?

5. What is the relationship of x to y?

6. Run each of the above programs for N = 10, 25, 40, and 55. Note for each the shape of the histogram as the value of N increases. Repeat if necessary. If you do not see a definite pattern emerging as the value of N gets larger, run the programs for N = 75 and N = 100. (This may take a while!)

7. What theorem in statistics explains the different shapes of the histograms obtained from these two programs?

Another application that is appropriate in an introductory statistics course is the generation of random numbers from a Poisson distribution. The following two programs for the HP-28S or HP48, POIS and RPOI, may be used for the generation and accumulation of 50 values that are chosen *at random* from a Poisson distribution with mean of λ. Both programs should be placed in the same directory.

Program: POIS[1] *for all HP users*

Purpose: To return one random value from a Poisson distribution with mean λ

Input: Input from the program RPOI is λ, the mean of the Poisson distribution

PROGRAMMING

```
<< NEG EXP -1 1 DO SWAP 1 + SWAP RAND *
   UNTIL DUP 4 PICK ≤ END DROP SWAP DROP
>>
```

Program: RPOI *for HP-28S users*

Purpose: Generate the random numbers and store them in ΣDAT.

Input: Input is the value of λ from within the program

PROGRAMMING

```
<< CLEAR Ø DUP → ℓ
   << { STO ℓ } MENU HALT CLΣ 1 50
      START ℓ POIS NEXT >>
   { 50 1 } →ARRY STOΣ
>>
```

- The character ℓ is obtained by pressing [LC] L.

[1]Wickes, William C., *HP-28 Insights: Principles and Programming of the HP-28C/S*, Larken Publications, Corvallis, Oregon, 1988.

Program: RPOI *for HP48 users*

Purpose: Generate the random numbers and store them in ΣDAT.

Input: Input is the value of λ from within the program

PROGRAMMING

```
<<  ERASE  Ø  DUP  →  λ
    <<  {  λ  }   MENU   HALT   CLΣ  1   50
          START   λ   POIS   NEXT  >>
    {  50   1  }   →ARRY   STOΣ
>>
```

HP users also have the ability to "seed" the random number generator RAND. Before using program RPOI, key in a value and press $\boxed{\text{RDZ}}^{M}$ which is found in the same menu as RAND. For instance, .2356 $\boxed{\text{RDZ}}^{M}$ $\boxed{\text{RPOI}}^{M}$ would yield different results than would .7845 $\boxed{\text{RDZ}}^{M}$ $\boxed{\text{RPOI}}^{M}$.

The 50 random numbers from the Poisson distribution with a user specified mean of λ are generated by POIS and stored into the statistics matrix ΣDAT with the keystroke $\boxed{\text{RPOI}}^{M}$. POIS is used as a subroutine of RPOI ("run" Poisson or "random" Poisson) and you should press only $\boxed{\text{RPOI}}^{M}$ to generate the random numbers. Summary statistics for the generated data may be obtained from the statistics menu of the calculator using the ΣDAT in the menu in which these programs reside. You may change the number of values generated by specifying a value other than 50 in the program RPOI.

HP-28S users may draw a histogram of the results by activating the programs →BINS and HIST. (If the programs POIS and RPOI are not in the same directory with the histogram programs, you must first transfer the matrix ΣDAT to the directory in which the histogram programs reside.)

HP48 users may draw a histogram of the results by activating the programs BINS and BARPL or HISTPL.

Random numbers from a normal distribution with a specified mean and variance may be generated using the uniform (pseudo) random number generator of the calculator with a definition of the random variable y as

$$y = \sqrt{-2 \ln x_i} \cos(2 \pi x_j)$$

where x_i and x_j are randomly drawn from the population of x. The variable y is then from a population conforming to the standard normal distribution. The distribution for a variable with mean $\bar{y}$ and standard deviation σ is then the normal distribution.[1] The following programs for the HP-28S or HP48 will generate numbers at random from the normal distribution. You may obtain a visual inspection of the distribution by drawing a histogram. Why not compare the results with those obtained from the randomly generated Poisson distribution?

All three programs that follow should be placed in the same directory.

Program: NORM *for all HP users*

Purpose: To return one random value from the standard normal distribution

Input: None

PROGRAMMING

```
<< RAND LN -2 * √
   RAND 2 * π →NUM
   * RAD COS *
>>
```

[1]Wickes, William C., *HP48 Insights: Part I* , Larken Publications, Corvallis, Oregon, 1991.

Program: MNORM *for all HP users*

Purpose: To return one value from a normal distribution with mean $\overline{y}$ and standard deviation σ

Input: None

PROGRAMMING

```
<< NORM * + >>
```

Program: RNOR *for HP-28S users*

Purpose: Generate the random numbers and store them in ΣDAT.

Input: Input from within the program are the desired values of the mean (M) and the standard deviation (S) of the normal distribution

PROGRAMMING

```
<< CLEAR { STO M S } MENU HALT CLΣ
   1 50 START M S MNORM NEXT
   { 50 1 } →ARRY STOΣ
>>
```

Program: RNOR *for HP48 users*

Purpose: Generate the random numbers and store them in ΣDAT.

Input: Input from within the program are the desired values of the mean (μ) and the standard deviation (σ) of the normal distribution

PROGRAMMING

```
<< { μ σ } MENU HALT CLΣ
   1 50 START μ σ MNORM NEXT
   { 50 1 } →ARRY STOΣ
>>
```

One random number from the normal distribution is generated by NORM, the mean and standard deviation of the distribution are specified through program MNORM, and the 50 values from that normal distribution are stored into the statistics matrix ΣDAT with the keystroke [RNOR]M. NORM and MNORM are used as subroutines of RNOR (normal "run" or normal "random") and you should press only [RNOR]M to generate the random numbers. Summary statistics for the generated data may be obtained from the statistics menu of the calculator using the ΣDAT in the menu in which these programs reside. You may change the number of values generated by specifying a value other than 50 in program RNOR.

HP-28S users may draw a histogram of the results by activating the programs →BINS and HIST. (If the programs NORM, MNORM and RNOR are not in the same directory with the histogram programs, you must first transfer the matrix ΣDAT to the directory in which the histogram programs reside.) HP48 users may draw a histogram of the results by activating the programs BINS and BARPL or HISTPL. Refer to the section on drawing histograms in Chapter 11 of this manual if you need assistance with the procedure.

SOLVING TECHNIQUES

Formulas may be entered on the stack, appropriate values entered for the variables in the formula, and the equation solving application of your Hewlett-Packard calculator can be used to solve for the variable in question. (HP48 users may enter formulas on the stack or with the EquationWriter application.) You will again find it helpful to create directories by application area to contain any programs you enter.

SOLVING FOR DISCRETE PROBABILITIES

For instance, suppose you wish to find the binomial probability that $x = 15$ for $n = 26$ and $p = .68$. Most statistics texts do not have binomial tables including these values of n and p so you will find your calculator quite useful. Since you will probably be finding many binomial probabilities, it is helpful to enter the equation of the binomial probability distribution in a program so that it can be used again with a minimum of key presses. For example, you could enter

'COMB(N,X)* P y^x X * (1-P) y^x (N-X)'

and then go the the SOLVE menu, store the equation with STEQ , enter the solver application with SOLVR and solve for the expression after you enter the values of n, p, and x. When you return to your user menu, you will see that menu keys for N, P, X, and EQ have appeared. If you are going to enter several formulas in the directory containing this program, the menu will get so cluttered that it will be cumbersome to use. The following program makes use of several nice features of the calculator that will eliminate some of the keystrokes necessary in solving and will eliminate the clutter from your user menu. Menus are accessed from within the program. HP-28S users will find a list of menu numbers on page 142 of the *Reference Manual* and HP48 users will find a list of menu numbers on page 697 of Volume II of the *Owner's Manual*.

The following program BINO should be entered and stored in your DSCR (discrete probability applications) directory.

Program BINO *for HP-28S users*

PROGRAMMING

```
<<  CLEAR  ' P ^ X * (1 - P ) ^ ( N - X ) * C '  STEQ
    24 MENU  HALT  N  X  COMB  'C'  STO  HALT
    23  MENU  { C  X  N  P  EQ }  PURGE
>>
```

EXPLANATION OF PROGRAMMING

- The command CLEAR clears the stack.
- The command STEQ stores the equation in your program as the current equation to be used by the SOLVR application.
- The command 24 MENU places you in the SOLVR application of the SOLVE menu.
- The first HALT command is used in order that you may have the calculator find the value of C = comb(N,X) and store it as the value of C for use by the equation. This is necessary on the HP-28S since it does not allow the function COMB to be included in an algebraic expression (' '). Notice the annunciator "o" appearing at the top left of the screen to tell you the program has halted for input of the values of N and X. The program is resumed and COMB(N,X) is calculated and stored in variable C by pressing ■ [CONT].
- The second HALT command tells the program to wait for input from the SOLVR application. Notice again that the annunciator "o" appears at the top left of the screen to tell you the program has halted for input of the value of P and execution of [EXPR=]M. (If you wish, the value of P may

be input with the values of N and X after the first HALT command.) The program is resumed after you have finished solving by pressing ■ [CONT].

- The command 23 MENU returns you to the USER menu in which you were last working; in this case, the DSCR menu.

- The keystrokes { C X N P EQ } PURGE eliminate the equation and variables used in the solving process from the DSCR menu after the answer has been found.

Program BINO *for HP48 users*

PROGRAMMING

```
<<  CLEAR  ' COMB ( N , X ) * P yX X * (1 - P ) yX
    ( N - X ) '  STEQ  3Ø  MENU  HALT  Ø  MENU
    { X N P EQ }  PURGE
>>
```

EXPLANATION OF PROGRAMMING

- The command CLEAR clears the stack.

- The command STEQ stores the equation in your program as the current equation to be used by the SOLVR application.

- The command 30 MENU places you in the SOLVR application of the SOLVE menu.

- The command HALT tells the program to wait for input from the SOLVR application. Notice that the word "HALT" appears at the top of the display screen when you are in the SOLVR application. The program is resumed after you have finished solving by entering the values of N, X, and P and pressing [EXPR=]M and then pressing [⬑] [CONT].

- The command 0 MENU returns you to the last menu you were using at the start of the program; in this case, the DSCR menu.
- The keystrokes { X N P EQ } PURGE eliminate the equation and variables used in the solving process from the DSCR menu after the answer has been found.

Values are input once you are in the SOLVR application by keying in the numerical value and pressing the menu key corresponding to that value. The equation is solved by pressing [EXPR=]M.

Run the program for our example with n = 26, x = 15 and p = .68. You will probably obtain the answer 8.55553455604E-2. Press [MODE]. Do you see [STD•]M? If so, this answer appeared as it did because scientific notation was used because of all the decimal places. If you wish the answer to appear to 4 decimal places, press 4 [FIX]M and you will see .8556. Note that this setting for 4 decimal places will remain in effect for display of all calculations until it is manually changed.

❐ CONFIDENCE INTERVALS

Because of the built-in equation solving application and upper-tail probabilities of your Hewlett-Packard calculator, it is not necessary to enter as complete programs all the formulas you will use in your statistics course. HP users should refer to the appropriate chapters in this manual for the concepts and conditions for use of the various confidence interval and hypothesis testing formulas in an introductory statistics course. HP 28-S users should read pages 254-257 in the *Reference Manual* and HP48 users should also read the section entitled Test Statistics, pages 383-385, in Volume I of the *Owner's Manual*.

The following programs, NPRB and NVAL, will allow you to use the built-in upper tail probabilities for the normal distribution as applied to hypothesis testing, confidence intervals and p-values (observed significance level).

Program NPRB[1]

PROGRAMMING

```
<< << μ σSQ X UTPN >> STEQ 3Ø MENU >>
```

This program will place you in the SOLVR application and place the variables for the mean and variance of the normal distribution and the value of X to be used in the SOLVR menu. Execute the program with NPRB[M] and enter the values of the variables. Press EXPR=[M] and you will have returned the upper-tail normal probability for the entered value of X. For instance, suppose you wish P(z > 1) for the standard normal distribution. Enter μ = 0, σSQ = 1 and X = 1. Press EXPR=[M] and you will see EXPR: 0.1587.

- If you wish, you may modify this program and the following one using the methods of the previous section so that the variables μ, σSQ, X, and EQ will not appear in your user menu.

Program NVAL

PROGRAMMING

```
<< << P μ σSQ X UTPN - >> STEQ 3Ø MENU >>
```

---

[1]These programs are given for the HP48. The only changes for HP-28S users are that the variables μ and σSQ will have to be replaced by M and V and that 30 MENU is replaced by 24 MENU.

This program will place you in the SOLVR application and place the variables for the mean and variance of the normal distribution and the values of X and P to be used in the SOLVR menu. This program is designed so that P equals the probability that the normal variable is *greater than or equal to* X. Execute the program with [NVAL]^M and enter the values of the variables P, μ, and σSQ. Press [↰] [X]^M to solve for X. For instance, suppose you wish the value of z, say z_0, such that $P(z > z_0) = 0.025$. Enter $\mu = 0$, σSQ = 1 and P = .025. Press [↰] [X]^M and you will see X: 1.9600.

If you wish to find a confidence interval for the mean μ of a population where it is appropriate to use the z test statistic, the program NVAL will give you the lower and upper endpoints of the confidence interval without you having to use any formulas or tables in your text!

Consider the problem: "A random sample of 80 observations from a population yielded a sample mean of 14.1 and a standard deviation of 2.6. Find a 95% confidence interval for the mean of this population." You must remember that the mean of the sampling distribution of the sample mean, $\mu_{\bar{x}}$, equals the population mean μ and that the standard deviation of the sampling distribution of the sample mean, $\sigma_{\bar{x}}$, equals $\sigma/\sqrt{n}$. (Refer to page 115 of this manual.) Use program [NVAL]^M and enter 14.1 [μ] 2.6 [ENTER] 80 [ENTER] [√] [÷] [x^2] [σSQ] .975 [P]. Press [↰] [X] to solve for X and you will see the lower limit of the confidence interval as X: 13.5303. Key in .025 [P] and press [↰] [X] to solve for X and you will see the lower limit of the confidence interval as X: 14.6697 .

- The values you have entered will remain stored in those variables until you store new values or use the SOLV application with another program. If you wish to see what values have been stored in the variables, return to your user menu and press the corresponding key.

- If you do not wish the variables and tools used by the solver application to clutter your menu, insert { X σSQ μ P EQ } PURGE at the end of program NVAL.

Confidence intervals are quite easily calculated from raw data that has been entered in the ΣDAT matrix using the following program. The following program makes use of the built-in upper tail normal probabilities of the HP48 calculator to give a large sample confidence interval . Raw data should be entered in the matrix in the statistics mode. Before the STAT mode is accessed to enter the data, be certain you are in the same directory in which the following program resides. If not, you will have to store the ΣDAT matrix in that directory before you can execute this program. Input from the stack on level 1 is $(1-\alpha/2)$ to determine the lower confidence limit and $\alpha/2$ to determine the upper confidence limit for a $(1-\alpha)(100\%)$ confidence interval.

Program Z ci μ *for all HP Users*

PROGRAMMING

```
<< 'P' STO MEAN 'M' STO SDEV SQ 'V' STO
   << P M V X UTPN - >> 'X' M ROOT
   { X P V M } PURGE
>>
```

- Input from the stack is P(x > L) where x is the normal random variable and L is the desired endpoint of the confidence interval. Data must be in ΣDAT.

Confidence intervals using the t-distribution for one-sample tests are programmed and used in a similar manner except that the number of degrees of freedom must be used in the program and you will use UTPT instead of UTPN. The following program, TVAL, is quite similar to program NVAL except that the Student t distribution is used instead of the normal distribution.

Program TVAL[1]

PROGRAMMING

```
<< << P f X UTPT - >> STEQ 3∅ MENU
HALT ∅ MENU { X f P EQ } PURGE >>
```

This program will place you in the SOLVR application and place the variables for the degrees of freedom (f) and the values of X and P to be used in the SOLVR menu. This program is designed so that P equals the probability that the t variable is *greater than or equal to* X. Execute the program with [TVAL] M and enter the values of the variables P and f. Press [↰] [X] M to solve for X. For instance, suppose you wish the value of t, say t_0, such that $P(t > t_0) = 0.025$ for 8 degrees of freedom. Enter P = .025 and f = 8. Press [↰] [X] M and you will see X: 2.306.

It is important to note that the function UTPT requires that the variable X is standardized. Thus, if you wish to use this program to determine upper and lower small sample confidence limits, you must first standardize the variable. The following program, T ci μ, is used in such cases. It is given below for the HP48. HP-28S users should make the appropriate menu changes and refer to previous examples for creating temporary menus of variables.

---

[1]This program is given for the HP48. The only changes for HP-28S users are that the 30 MENU is replaced by 24 MENU and ∅ MENU is replaced by 23 MENU.

The extra capabilities of the HP48 make it possible for you to name the variables with, in most cases, the same symbols as those appearing in the formulas in your textbook. The symbol $\bar{x}$ is obtained on the HP48 by pressing [α] [↱] [1/X] .

Program T ci μ

PROGRAMMING

```
<< 'P' STO { x̄ s n } MENU HALT n 1 - 'F'
   STO << P F X UTPT - >> 'X' 2 ROOT
   DUP IF 0 < THEN NEG END 'R' STO x̄
   s n √ / R * DUP2 - 3 ROLLD +
   { X P F R x̄ n s CST } PURGE Ø MENU
>>
```

- Input from the stack is $\alpha/2$ to yield the upper and lower confidence limits for a $(1-\alpha)(100\%)$ confidence interval. The values of the sample mean, the sample standard deviation and the sample size are input from within the program.
- $\bar{x}$ is the mean of the sample, s is the sample standard deviation and n stands for the sample size. The number of degrees of freedom, $n - 1$, is calculated from within the program.

When you are running program T ci μ, the program will halt for you to input the sample mean, sample standard deviation and sample size. After these values have been entered, press [↰] [CONT] . Refer to page 128 of this manual for the conditions under which this formula should be used. Exercises are given on page 129. Your answers may differ slightly because you are not using rounded values from the t table in your text.

For confidence intervals involving two-samples, you may find it easier to program the standard deviation of the test statistic as a separate quantity to avoid length stack calculations. Of course, if you wish, the formulas that are in your text may be programmed as algebraic expressions and solved using the SOLVR application. Several different methods of programming have been presented in this section. Refer to these when you are writing your own programs for evaluating the other confidence interval formulas in your text.

❑ HYPOTHESIS TESTING

If you are familiar with stack manipulations and RPN (Reverse Polish Notation), you can readily obtain solutions to most of the "simpler" equations. You can also use the programming capabilities of the calculator to calculate the value of the test statistic. Suppose you are testing a hypothesis concerning the mean of a single population H_0: $\mu = 12$ versus the alternative H_a: $\mu > 12$ and the conditions are such that you should use the z test statistic. Enter the following program:

Program ZHTM *for HP-28S users*

PROGRAMMING

```
<<  CLEAR  ' ( XB - M ) * √ N / S ' STEQ  24  MENU
       HALT  23  MENU  { XB  M  N  S  EQ }  PURGE
>>
```

- The symbol XB refers to $\bar{x}$, the mean of the sample, the symbol M stands for μ, the mean of the population, N stands for the sample size, and S is used for the population standard deviation σ.

Program **Z ht μ** *for HP48 users*

PROGRAMMING

```
<< CLEAR '(x̄ - μ) * √ n / σ' STEQ 3Ø MENU
   HALT Ø MENU { x̄ μ n σ EQ } PURGE
>>
```

- $\bar{x}$ is the mean of the sample, μ is the mean of the population, n stands for the sample size, and σ is the population standard deviation.

For the hypothesis test H_o: $\mu = 12$ versus the alternative H_a: $\mu > 12$, suppose we are given that the mean of a random sample of size 45 chosen from the population of interest is 12.92 and the sample standard deviation is 5.95. Execute ZHTM with the following values: $\bar{x} = 12.92$, $\mu = 12$, $\sigma = 5.95$ (use s to approximate σ due to the large sample) and n = 45. You will find EXPR = z = 1.0372. This value can be compared to the critical value for z based on a given value of α = P(Type I error) and the appropriate conclusion reached.

If you wish to determine observed significance level (p-value) for a hypothesis test, you can use the programs NPRB and NVAL or ones similar to these for the appropriate test statistic. Let's find the p-value for the hypothesis test conducted above. Now, the p-value = P(z > 1.0372) can be obtained by two different methods:

1) Using the program NPRB, substitute, for the standard normal (z) distribution, $\mu = 0$, σSQ = 1, X = 1.0372 and solve for the probability with [EXPR=]M. You will obtain an answer of 0.1498 = P(z > 1.0372). Thus, the null hypothesis will be rejected for any value of $\alpha > .1498$.

2) Using the program NVAL, substitute $\mu = 12$, σSQ = $(5.95 / \sqrt{45})^2$, X = 12.92 and solve for the probability with [↰] [P]M. You will obtain an answer of 0.1498 = P($\bar{x}$ > 12.92). Thus, the null hypothesis will be rejected for any value of $\alpha > .1498$.

EXERCISES

1. Test the hypothesis H_0: $\mu = 8$ against the alternative H_a: $\mu > 8$ using the following information from a random sample of size 36: the sample mean is 9.35 and the sample standard deviation is 4.56. The level of significance for the test is chosen to be 0.05. Find the p-value for this hypothesis test.

2. A random sample of size 100 yielded a mean of 21.45 and a standard deviation of 15.82. Using P(Type I error) = .05, test H_0: $\mu = 23.5$ against the alternative hypothesis H_a: $\mu \neq 23.5$. Find the p-value for this hypothesis test.

ANSWERS

1. z = 1.7763 so reject H_0 p-value = .0378
2. z = - 1.2958 so do not reject H_0 p-value = 2(.0975) = .1950.

Note that if you use the program NVAL to find the p-value in problem 2, you will obtain an answer of (probability) P = .9025. Recall that the probability returned by this program is P(normal distribution x ≥ x value from the sample). Because the sample mean value is less than the population mean value (z is negative), the p-value = 2 P ($\bar{x} < 21.45$) = 2 (1 – .9025) = .1950. The reason you get .9025 as the value of the expression in the program NPRB when -1.2958 is substituted for X is the same. If you use X = 1.2958 in the program NPRB, you will obtain the expression equal to .0975. The multiplication by 2 is because this is a two-tail hypothesis test.

Refer to the exercises in Chapters 7 and 8 of this manual for more practice on confidence intervals and hypothesis testing techniques.

❏ EXTENSIONS

Upper-tail probabilities for the Chi square distribution and F distribution are also built into your calculator and used as explained above for the upper-tail normal probabilities.

When you are given raw data instead of summary statistics, you can enter the data into the statistical matrix ΣDAT with the Σ+ key and use the STAT menu to obtain the mean and standard deviation of the data. You can then use the formulas that ask for summary statistics. If you are ambitious, try programming your own formulas, such as Z ci μ, that use raw data .

For regression analysis you will find your HP calculator quite a marvelous tool. Scatter diagrams and regression models are quite easily graphed. The HP48 even has a menu key that will choose the best-fitting model for your data!

Explore and experiment as you use your calculator in your statistics course and you will find many more applications that will aid you in the computational aspects of the course and the understanding of the concepts presented in class. Good luck!

APPENDIX I

The following gives an explanation of the algebraic manipulation and programming on the Sharp EL-5200 for BINOMIAL 7Ø:

Recall that the binomial probability formula is $p(x) = \binom{n}{x} p^x (1-p)^{n-x}$.

$$\text{Now, } \binom{n}{x} = nCx = \frac{n!}{x!\,(n-x)!} = \frac{n(n-1)(n-2)\ldots(n-x+1)(n-x)!}{x(x-1)(x-2)\ldots(3)(2)(1)(n-x)!} =$$

$$\frac{n(n-1)(n-2)\ldots(n-x+1)}{x(x-1)(x-2)\ldots(3)(2)(1)} = \frac{(n-x+1)\ldots(n-2)(n-1)(n)}{x(x-1)(x-2)\ldots(3)(2)(1)} =$$

$$\frac{(n-x+1)(n-x+2)(n-x+3)\ldots(n-1)(n)}{x(x-1)(x-2)\ldots(3)(2)(1)} =$$

$$\frac{[n-(x-1)][n-(x-2)](n-(x-3)]\ldots(n-1)(n)}{x(x-1)(x-2)\ldots(3)(2)(1)} =$$

$$\left(\frac{n-(x-1)}{x}\right)\left(\frac{n-(x-2)}{x-1}\right)\left(\frac{n-(x-3)}{x-2}\right)\ldots\left(\frac{n-1}{2}\right)\left(\frac{n}{1}\right) =$$

$$\left(\frac{n+1-x}{x}\right)\left(\frac{n+1-(x-1)}{x-1}\right)\left(\frac{n+1-(x-2)}{x-2}\right)\ldots\left(\frac{n+1-2}{2}\right)\left(\frac{n+1-1}{1}\right) =$$

$$\left(\frac{n+1}{x}-1\right)\left(\frac{n+1}{x-1}-1\right)\left(\frac{n+1}{x-2}-1\right)\cdots\left(\frac{n+1}{2}-1\right)\left(\frac{n+1}{1}-1\right)=$$

$$\left(\frac{n+1}{1}-1\right)\left(\frac{n+1}{2}-1\right)\left(\frac{n+1}{3}-1\right)\cdots\left(\frac{n+1}{x-1}-1\right)\left(\frac{n+1}{x}-1\right).$$

Notice that nCx, which initially involved the division of factorials, has been rewritten as a product of expressions, each involving one division without factorials. For $x = 0$ or $x = n$, $nCx = 1$. The algebracially simplified expression on the previous page results in undefined expressions when $x = 0$. Thus, the program BINOMIAL 7Ø checks to see if $x = 0$ or $x = n$ and sets $C = 1$ in these cases.

In the flowchart on the next page, D refers to the denominator of the above expression which is set to its initial value of 1. C, which is also initially set to 1, first computes (by the formula given in the flowchart) the value of the leftmost product. D is then increased by 1 and the value of D is checked to see if the entire product has been computed. If not, the next term in the product is picked up and multiplied by the previous value. This process continues until the desired result is reached and substituted in the expression for the binomial probability.

PROGRAM FLOW CHART

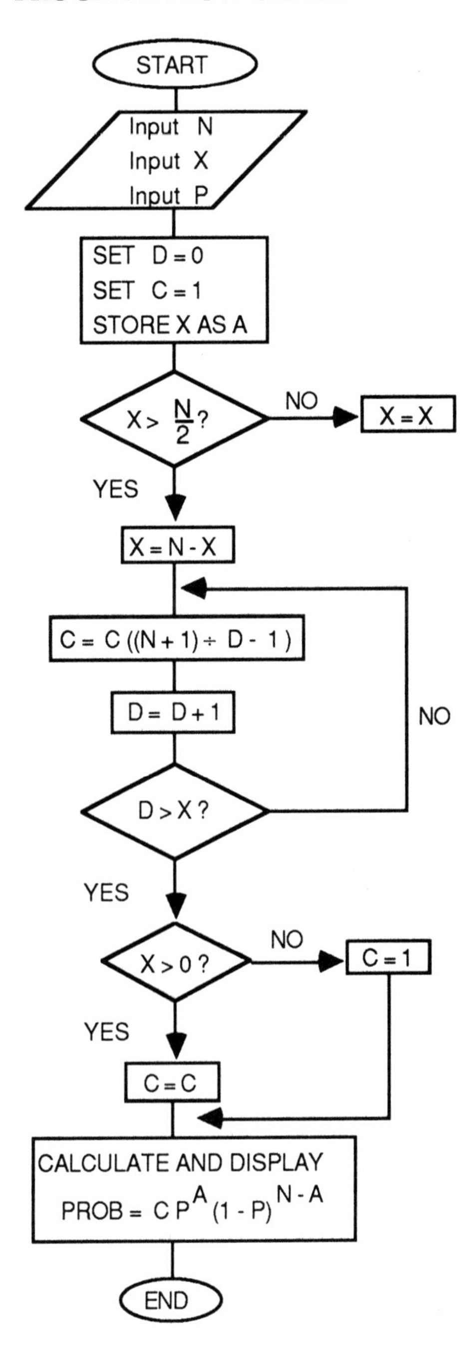

PROGRAM INDEX I

(by application area)

DESCRIPTIVE STATISTICS

---

[1]Programs of the same name are included for both the HP-28C and the HP48.

GENERAL UTILITY

DISCRETE PROBABILITY DISTRIBUTIONS

SAMPLE SIZE

SIMULATION

CONTINUOUS PROBABILITY DISTRIBUTIONS

CONFIDENCE INTERVALS

HYPOTHESIS TESTING

PROGRAM INDEX II

(by order of entry)

PROGRAMS FOR THE SHARP EL-5200

PROGRAMS FOR THE HP-28S

PROGRAMS FOR THE HP48

SUBJECT INDEX